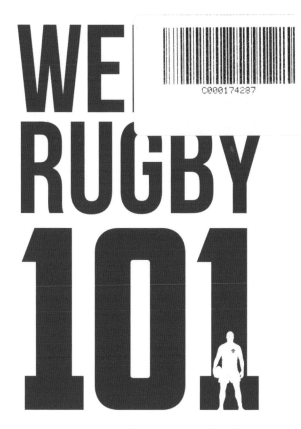

WE RUGBY 101

A POCKET GUIDE

IN 101 MOMENTS, STATS, CHARACTERS AND GAMES

JOHN GRIFFITHS

POLARIS
PUBLISHING

This edition first published in 2019 by

POLARIS PUBLISHING LTD
c/o Aberdein Considine
2nd Floor, Elder House
Multrees Walk
Edinburgh, EH1 3DX

Distributed by
ARENA SPORT
An imprint of Birlinn Limited

www.polarispublishing.com
www.arenasportbooks.co.uk

Text copyright © John Griffiths, 2019

ISBN: 9781909715790
eBook ISBN: 9781788851800

British Library Cataloguing-in-Publication Data
A catalogue record for this book is available on request from the British Library.

Designed and typeset by Polaris Publishing, Edinburgh

Printed in Great Britain by MBM Print SCS Limited, East Kilbride

Photos courtesy of:
Inphophotography
Getty Images
John Griffiths' archive
Arena Sport archive

MIX
Paper from responsible sources
FSC® C117931

Acknowledgements

I am indebted to John Jenkins, Howard Evans and Tony Lewis, three longstanding Welsh friends, for making freely available the fruits of their research into Welsh rugby. Cuttings from the *Western Mail*, Wales's national daily, have also been a constant source of reference and I am indebted to their former rugby correspondents, notably John Billot, JBG Thomas and WJ Hoare for the accuracy with which they reported Welsh international rugby. But above all, thank you to the 1151 players capped by Wales since 1881: without them there would be no book.

Introduction

Welsh rugby fans have never had it so good. As this pocket guide goes to press, the national side is enjoying a record run of successive victories and occupies its highest standing, to date, in the official rankings calculated by World Rugby, the sport's governing body.

So is the class of 2019 the best Welsh team of all time? That debate will be a part of rugby chat in pubs and clubs throughout the Principality as Warren Gatland's team prepares for the Rugby World Cup in Japan later this year.

As the only perspective on the present is the past, this collection might help inform the arguments. In a distillation of 101 facts and stats, famous characters and matches, I have set out to capture the achievements and some of the disappointments of the national side from its modest entry to Test rugby in February 1881 to the joy of its recent Grand Slam triumph in Cardiff in March 2019.

The choice of what to include is mine and inevitably others will have different views. 'How could you leave out Percy Bush or Dai Watkins?' will be the cry from Cardiff and Newport. Or the million who claim to have been there might ask why accounts of Wales beating England at Wembley in 1999 and Cardiff in 2013 are omitted. Readers can suggest alternatives, but in the meantime here is my selection of the 101 men, matches and moments that stand out from more than 138 years of Welsh rugby.

The statistics go up to 31st March, 2019

FAMOUS FIRSTS

The people's game

It was probably near Newport (Mon) that the Romans, during their western raids, first unveiled to the people of South Wales a game that resembled rugby. Then, in medieval times, West Wales was a stronghold of the mob game known as cnapan, which featured handling and kicking between two teams of mauling, brawling masses and which, with modifications, survived into the early 19th century. So when the Rugby School code of football was introduced to South Wales by students, teachers and doctors returning from English institutions where the game was practised, its distinctive feature of running with the ball struck a chord with young Welshmen.

Rugby football had established a foothold in Welsh schools and colleges before, in 1866, the first competitive match on Welsh soil was staged at Lampeter in rural West Wales, the participants being the local St David's College and Llandovery College, Wales's leading public school. Welsh scholars departing their cloistered confines subsequently assumed the lead role in spreading the game. The rapid industrialisation of South Wales, moreover, was the driving force of a population explosion that brought English, Scots and Irish to the area to support the native workforce in the coalfields and furnaces. Many joined with students and clerical workers supporting the heavy industries to establish the rugby clubs that sprouted in the 1870s, helping rugby football become the game of the Welsh people by the end of the decade.

FAMOUS FIRSTS

An Aussie captain

Newport were the leading Welsh rugby club of the late 1870s and early 1880s and it was a Newport initiative taken by their ambitious secretary, Richard Mullock, that led to Wales placing her first international fifteen on the field. Newport were affiliated to the Rugby Football Union and Mullock persuaded the RFU's grandees to grant the Principality a fixture with England for the 1880/81 season. Mullock took it upon himself to raise the team.

The first Welsh captain chosen by Mullock was Australian-born James Alfred Bevan, who stares out from the team group photo for that encounter with England at Blackheath in 1881 wearing the red jersey and Prince of Wales feathers that became the permanent emblem of Welsh rugby fifteens. Bevan had first seen the light of day in St Kilda in Victoria on 15th April 1858 but was sent to the Welsh marches to be educated at Hereford Cathedral School after he was orphaned as a child. The future captain was a gifted sportsman who excelled at cricket, golf and rugby, winning Cambridge Blues in 1877 and 1880. But his single match in charge of the Welsh rugby team proved an utter disaster, the

Principality suffering a humiliating defeat by seven goals, six tries and a dropped goal to nil – an 82–0 hiding in today's money.

Leading the dragon

All told, 137 men have worn the captain's armband since Wales made their inauspicious entry to international rugby in 1881.

Name	First captaincy		
JA Bevan	19 Feb 1881	RT Gabe	9 Mar 1907
CP Lewis	28 Jan 1882	AF Harding	18 Jan 1908
HJ Simpson	12 Apr 1884	G Travers	1 Feb 1908
CH Newman	3 Jan 1885	E Morgan	2 Mar 1908
FE Hancock	9 Jan 1886	HB Winfield	14 Mar 1908
R Gould	26 Feb 1887	RA Gibbs	12 Mar 1910
TJS Clapp	12 Mar 1887	JL Williams	28 Feb 1911
AF Hill	22 Dec 1888	J Bancroft	9 Mar 1912
AJ Gould	2 Mar 1889	TH Vile	25 Mar 1912
WA Bowen	3 Jan 1891	JP Jones	8 Mar 1913
WH Thomas	7 Feb 1891	Rev JA Davies	17 Jan 1914
WJ Bancroft	18 Mar 1898	G Stephens	21 Apr 1919
EG Nicholls	11 Jan 1902	H Uzzell	17 Jan 1920
TWR Pearson	10 Jan 1903	JJ Wetter	15 Jan 1921
GL Lloyd	7 Feb 1903	T Parker	26 Feb 1922
W Llewellyn	6 Feb 1904	JMC Lewis	20 Jan 1923
RM Owen	12 Jan 1907	A Jenkins	10 Mar 1923
WJ Trew	2 Feb 1907	J Rees	19 Jan 1924
		J Whitfield	2 Feb 1924

R Harding	27 Mar 1924	AEI Pask	15 Jan 1966
TAW Johnson	17 Jan 1925	D Watkins	11 Mar 1967
S Morris	7 Feb 1925	NR Gale	11 Nov 1967
RA Cornish	28 Feb 1925	GO Edwards	3 Feb 1968
WI Jones	14 Mar 1925	SJ Dawes	9 Mar 1968
WJ Delahay	5 Apr 1926	B Price	1 Feb 1969
BR Turnbull	15 Jan 1927	DJ Lloyd	15 Jan 1972
BO Male	5 Feb 1927	WD Thomas	2 Dec 1972
WC Powell	26 Feb 1927	AJL Lewis	20 Jan 1973
IE Jones	26 Nov 1927	TM Davies	18 Jan 1975
WG Morgan	2 Feb 1929	P Bennett	15 Jan 1977
HM Bowcott	18 Jan 1930	TJ Cobner	11 Jun 1978
JA Bassett	8 Mar 1930	TGR Davies	17 Jun 1978
WG Thomas	21 Jan 1933	JPR Williams	11 Nov 1978
JR Evans	20 Jan 1934	J Squire	19 Jan 1980
C Davey	3 Feb 1934	SP Fenwick	1 Nov 1980
JI Rees	18 Jan 1936	WG Davies	5 Dec 1981
W Wooller	3 Apr 1937	ET Butler	5 Feb 1983
CW Jones	15 Jan 1938	MJ Watkins	4 Feb 1984
H Tanner	18 Jan 1947	TD Holmes	2 Mar 1985
WE Tamplin	20 Dec 1947	DF Pickering	17 Jan 1986
JA Gwilliam	21 Jan 1950	RD Moriarty	12 Jun 1986
J Matthews	7 Apr 1951	WJ James	4 Apr 1987
BL Williams	7 Feb 1953	J Davies	3 Jun 1987
JRG Stephens	16 Jan 1954	B Bowen	7 Nov 1987
WR Willis	27 Mar 1954	RL Norster	28 May 1988
KJ Jones	10 Apr 1954	PH Thorburn	21 Jan 1989
CI Morgan	21 Jan 1956	RN Jones	4 Nov 1989
MC Thomas	19 Jan 1957	KH Phillips	2 Jun 1990
RCC Thomas	4 Jan 1958	IC Evans	4 Sep 1991
RH Williams	16 Jan 1960	GO Llewellyn	22 May 1993
BV Meredith	6 Feb 1960	MR Hall	27 May 1995
DO Brace	12 Mar 1960	JM Humphreys	2 Sep 1995
TJ Davies	3 Dec 1960	NG Davies	25 Sep 1996
LH Williams	25 Mar 1961	IS Gibbs	11 Jan 1997
DCT Rowlands	19 Jan 1963	RG Jones	5 Jul 1997

P John	19 Jul 1997	DJ Peel	9 Sep 2007
R Howley	7 Feb 1998	GD Jenkins	24 Nov 2007
KP Jones	27 Jun 1998	RP Jones	2 Feb 2008
D Young	5 Feb 2000	A-W Jones	14 Mar 2009
M Taylor	11 Nov 2000	M Rees	6 Nov 2010
LS Quinnell	26 Nov 2000	SK Warburton	4 Jun 2011
AP Moore	10 Jun 2001	BS Davies	8 Jun 2013
CL Charvis	6 Apr 2002	MS Williams	8 Aug 2015
ME Williams	8 Mar 2003	DJ Lydiate	19 Mar 2016
G Thomas	16 Aug 2003	JH Roberts	16 Jun 2017
SM Jones	23 Aug 2003	TT Faletau	11 Mar 2018
DWMM Davies	27 Aug 2003	EL Jenkins	2 Jun 2018
MJ Owen	13 Mar 2005	CL Hill	9 Jun 2018
DJ Jones	11 Jun 2006	JJV Davies	9 Feb 2019

The creation of the WRU

The humiliation of that first Welsh defeat at Blackheath, coupled with general dissatisfaction at the autocratic way in which Richard Mullock had set about organising and selecting the Welsh fifteen, provoked a backlash in Wales. At length, a meeting at the Castle Hotel, Neath, on 12th March 1881 brought together the great and the good of Welsh rugby. Nearly a dozen clubs, including Newport, Cardiff, Swansea, Llanelli and Neath, sent delegates and the culmination was the formal foundation of the Welsh Football Union (officially rebranded as the Welsh Rugby Union in 1935).

In what must have been a magnanimous gesture, the fledgling Union unanimously elected Richard Mullock as its secretary and entrusted him with selecting a fully representative team to meet Ireland in April – a fixture that eventually fell through, though Wales did play their inaugural international against Ireland in January 1882 and their first with Scotland in January 1883.

In 1886, the Welsh Union became founder members with their Scottish and Irish counterparts of the International Rugby Board, the sport's governing body which is better known today as World Rugby.

The other side of the whistle

Rugby's version of a poacher turned gamekeeper is the former international player who becomes a Test referee. It is very rare today – Ireland's Alain Rolland is the only example in recent decades – but it was more common in the early days of Test rugby. Nine Welsh internationals swapped playing at the highest level to whistle in international rugby.

Name	First Welsh Cap	First Test as Referee
WD Phillips	1881	1887
CP Lewis	1882	1885
T Williams	1882	1904
DH Bowen	1882	1905
J Griffin	1883	1891
HS Lyne	1883	1885
W Douglas	1886	1891
EG Nicholls	1896	1909
TH Vile	1908	1923

The Welsh system

Wales entered international rugby ten years behind England and Scotland, and six seasons after Ireland had made their debut. By 1881, passing among forwards and backs was just being explored at the fashionable English clubs. This benefited open play because the backs were released from their primary duties as defenders and converted into potential scoring forces. The Welsh clubs, riding the winds of change, felt that the time was ripe to experiment with back formations in order to capitalise on the passing game. First the back division was tweaked by redeploying one of the two full backs as a third threequarter, and then Cardiff (accidentally at first) began the practice of removing a player from the scrum to operate with eight forwards and seven backs. The extra man was positioned between the halves and the full back, giving birth to the four threequarter line, a Welsh innovation that transformed rugby into its modern format.

Cardiff's Frank Hancock was more than the accidental inspiration behind the new system. His playing and leadership credentials were well-known in rugby circles when he moved from his Wiveliscombe home to Cardiff in late 1883 as a director with the family brewery business. He had captained the Somerset club side through a long unbeaten home run and had led his county with success, so it was no surprise that within two years of his arrival in Cardiff he was elected club skipper. Under his direction, Cardiff played the four threequarter game with outstanding success in the 1885–86 season, winning 26 successive games and scoring 131 tries.

Hancock believed that the whole was greater than the sum of its parts and created a side that subscribed to his view that playing in combination rather than as individuals was more productive. Through example and force of personality he got more out of his team than any previous Cardiff captain had done. Hancock's ideas worked at club level: Cardiff's leading two try-scorers for the season were threequarters and, in January 1886, Hancock was rewarded with the captaincy of his adopted country for Wales's encounter with Scotland at Cardiff Arms Park. Wales, moreover, bowed to the Cardiff system by fielding four threequarters and, for the first time in an international, the pack was reduced to eight forwards.

Although Hancock carried five of his team-mates into the national fifteen, including the prolific runner-in of tries Billy Douglas who headed the Cardiff lists that season, what had worked at club level was a disaster against a heavy Scottish pack whose nine forwards constantly ran through the Welsh eight. So overwhelmed were the home side that at half-time Harry Bowen, who had taken the field at full back, was moved up to strengthen the pack, and Arthur Gould, the promising young Newport centre threequarter whose debut the previous season had been greeted with rave reviews, was forced to drop back and cover for Bowen. The young Gould was not impressed and for some time at the start of his long career as a Welsh player remained circumspect about the four threequarter game.

FAMOUS CHARACTERS

Welsh rugby's first legend

Arthur Gould was Welsh rugby's first legend, the sport's equivalent to cricket's Dr W G Grace and the game's most colourful player of the nineteenth century. As a teenaged prodigy Gould was noted for his exceptional pace, accurate kicking and fearless defence. He was first capped as a full back in 1885 but soon moved into the centre where his individual brilliance as the sole occupant in the middle of the three threequarter line quickly drew rave reviews from the critics. He was a natural ball player, but it was his adaptability and tactical insight that were to bring him superstar status. A formative period spent in England between 1886 and 1890 taught him the benefits of combined play, and when Wales regularly adopted the four threequarter system in the 1890s, Gould overcame his early resistance to the formation and became its expert exponent and fervent champion.

He captained Wales to their first Triple Crown in 1893, scoring two tries in the 12–11 victory over England and creating the overlap for his co-centre to cross for the only score of the match with Ireland. Wales's success under Gould prompted the other Home Unions to embrace the four threequarter game thereafter. In 1896, his admirers (including the Welsh Rugby Union) contributed to a testimonial fund, but the International Board regarded the move as an act of professionalism. Wales were banned from international matches until Gould took matters into his own hands by gracefully retiring in 1897 after making a then record 27 appearances for his country.

Living up to his nickname

As a youngster Arthur Gould had demonstrated agility as a tree-climber. He was so adept clambering up tree-trunks and swinging through branches that his young pals nicknamed him "Monkey". It was a moniker he carried into his playing days and, at Llanelli in January 1887, the crowd assembled for the Wales–England international saw him live up to his name. Stradey Park was so frost-bound that the match had to be staged on the adjacent cricket pitch, the local ground-staff hurriedly erecting a makeshift set of posts so that the fixture could go ahead. During the match one of the crossbars fell to the ground, whereupon Gould performed his old party piece, shinning up one of the uprights with alacrity to replace the fallen crossbar.

9

STAT ZONE

Leading skippers

The remarkable Arthur Gould led Wales 18 times in the days when the Home Unions played only three international matches a year. It was the Welsh captaincy record for nearly a hundred years and to this day Gould remains in the top ten of the Principality's leading skippers.

Tests	Captain	Captaincy span
49	SK Warburton	2011 to 2016
33	RP Jones	2008 to 2013
28	IC Evans	1991 to 1995
24	AW Jones	2009 to 2019
22	R Howley	1998 to 1999
22	CL Charvis	2002 to 2004
21	G Thomas	2003 to 2007
19	JM Humphreys	1995 to 2003
18	AJ Gould	1889 to 1897
14	DCT Rowlands	1963 to 1965
14	WJ Trew	1907 to 1913

FAMOUS FIRSTS

Living with Lions

Wales has provided British & Irish Lions touring teams with more than 150 players, 148 appearing in official Test matches against Australia, New Zealand, South Africa and Argentina. The first officially recognised tour was the visit to Australia and New Zealand in 1888 when William Henry Thomas, a native of Fishguard and a Cambridge Blue, had the honour of becoming Wales's first Lion. A forward, he was only 21 when he boarded the *Kaikoura* in Liverpool bound for the Antipodes, but was already a seasoned Welsh international having gained his first cap three years earlier as a teenager.

FAMOUS CHARACTERS

The confident Mr Bancroft

The most confident player of his generation: that was the epitaph of Billy Bancroft, the Swansea cobbler who monopolised the full back position in the Welsh team between 1890 and 1901. He made his debut against Scotland in 1890, entering the side only as a late replacement for the injured Newport full back, Tom England. Bancroft never missed a match in a dozen seasons, finally standing down with a remarkable record of 33 successive internationals in the days long before games were staged against France and overseas sides.

Bancroft was a little man who used his speed and keen sense of anticipation to outwit teams. He was the scourge of opposing forwards. Rather than returning the ball to touch after making a clean catch, he invariably paused to tease opposing packs to catch him. Then, as forwards advanced he would set off on long intricate tracks that frustrated and eventually wore them out. Meticulous preparation was the secret of Bancroft's success as a kicker. He practised for hours on the St Helen's ground in Swansea to perfect his technique. He could land goals from the corner flag, from halfway or even from one yard in front of the posts. During his long career he took every kick at goal that Wales were offered and his haul of 60 points stood as the Welsh Test record until passed by his younger brother, Jack, shortly before the Great War.

12

Beating England at Cardiff

WALES 12, ENGLAND 11
Cardiff Arms Park, 7 January 1893

Severe cold weather had threatened the match and only the efforts of the Cardiff groundsman and a gang of unemployed labourers, who burnt fire devils on the pitch throughout the eve of the match, ensured the fixture went ahead.

England opened strongly and led 7–0 at half-time. The visitors maintained the initiative after the interval with Howard Marshall, in his only appearance in an international match, finishing with a hat-trick of tries as England went out to a 11–9 lead with time running out.

Then came Wales's winning score. A penalty was awarded outside the English 25 but close to the touchline. The Welsh skipper ordered Billy Bancroft to place the goal, but the full back defied his captain and instead drop-kicked, sending the ball flying between the uprights. It was the first penalty goal ever landed in an international match and brought Wales a narrow 12–11 triumph on the first leg of a Triple Crown season.

FAMOUS FIRSTS

The Triple Crown

After registering their first Cardiff win against England in January 1893, Wales went on to beat Scotland comfortably (9–0) at Inverleith and finally clinched their first Triple Crown on the slopes of Llanelli's Stradey Park with a win against Ireland by a solitary try (scored by Arthur Gould's brother Bert) to nil.

All told, Wales have ruled the roost with a clean sweep of victories over the other Home Unions 21 times: 1893, 1900, 1902, 1905, 1908, 1909, 1911, 1950, 1952, 1965, 1969, 1971, 1976, 1977, 1978, 1979, 1988, 2005, 2008, 2012 and 2019.

14

FAMOUS CHARACTERS

The prince of centres

The player whose modernising effect transformed Welsh rugby at the turn of the twentieth century was Gwyn Nicholls, the man they called "The prince of centres."

Nicholls was an effective and elusive runner whose balance and skills took him into the Welsh XV in 1896. After that he was an integral part of a side that went on to enjoy a Golden Era in the early 1900s. He participated in three Triple Crown sides and led Wales to a famous victory against the 1905 All Blacks.

Nicholls's special contributions to success were exceptional judgement, shrewd philosophy and insightful captaincy. He firmly believed the effectiveness of a side was greater than the sum of its parts: "None is for a party, but all for the state," he wrote after he had retired.

For all his individual brilliance he never wavered from this belief, and for his principles he was held in the highest esteem. He was better remembered as the maker of tries for the great wings, Willie Llewellyn and Teddy Morgan, than as a prolific scorer. With these two wings and the Llanelli centre, Rhys Gabe, Nicholls formed the most potent threequarter line in world rugby. As a quartet they played together seven times and were never in a losing side.

His understanding of tactics guided Wales to the Triple Crown in his first season as captain, in 1902. All told he led his country in ten matches, winning seven. His final game was the defeat by the First Springboks in 1906, when Nicholls unwisely accepted an invitation to come out of retirement to lead Wales for the last time.

STAT ZONE

Most tries in a Test by a player

Willie Llewellyn was a household name during Welsh rugby's first golden era at the turn of the nineteenth century, crossing for 16 tries in 20 Tests between 1899 and 1905. The little Llwynypia wing launched his international career in memorable fashion, scoring four tries against England at Swansea in 1899. Many years later he modestly recalled: "As a youngster playing in my first match, I trembled as to what was likely to happen. However, I managed to score four tries and came off the field feeling very pleased with life. Chief reason for my success was the openings made for me." Only nine Welshmen down the years have matched Llewellyn's feat.

Tries	Player	Opponents	Venue	Year
4	W Llewellyn	England	Swansea	1899
4	RA Gibbs	France	Cardiff	1908
4	MCR Richards	England	Cardiff	1969
4	IC Evans	Canada	Invercargill	1987
4	NK Walker	Portugal	Lisbon	1994
4	G Thomas	Italy	Treviso	1999
4	SM Williams	Japan	Osaka	2001

| 4 | TGL Shanklin | Romania | Cardiff | 2004 |
| 4 | CL Charvis | Japan | Cardiff | 2004 |

Llewellyn's achievement is unique as he is the only player to do so on his international debut. Charvis is the only forward among the nine.

Willie Llewellyn

16

FAMOUS CHARACTERS

The Bullet

He measured just 5 foot 1½ inches and scaled barely 9½ stones, yet Dicky Owen went on to become the world's most-capped international rugby player, holding the record for nearly 20 years. But despite his lack of height, he was a courageous tackler who enjoyed toppling big forwards with taps around their ankles. His fast, accurate service revolutionised scrum half play in the early 1900s and he loved creating chances for his colleagues, but he primarily saw his role as the tactical hub of the team.

Owen masterminded many famous victories during a Golden Era for Welsh rugby. He entered the side in 1901 and in his second international, against England at Blackheath in 1902, gave notice of his innovative genius by tricking his opponent off-side at a scrum, thus presenting Wales with a match-winning penalty in the first leg of a Triple Crown season.

All told, he featured in five Triple Crown seasons and three Grand Slams, but it was his part in the win over the Original All Blacks of 1905 that earned him lasting fame. He conceived and executed the feint move and reverse pass that wrong-footed the New Zealanders to create the opening for Teddy Morgan's sprint and decisive score.

Yet his pioneering role as a tactician did not meet with universal approval. At Cardiff and Newport the view was that fly half was the chief play-maker's ideal position and there was invariably conflict when Owen's partner for Wales came from beyond his club, Swansea. None, however, can deny that little Dicky was the father of modern scrum half play.

17

FAMOUS GAMES

Champions of the world

WALES 3, NEW ZEALAND 0
Cardiff Arms Park, 16 December 1905

The Original All Blacks cut an impressive unbeaten path through the British Isles at the start of the 1905 season. Their triumphs were ascribed to superior fitness and skilful combined play, and were trumpeted as a wonderful advertisement for the outdoor way of life of the southern colonies. It was left to "Gallant Little Wales" to burst New Zealand's bubble of invincibility. In front of a partisan crowd of 40,000, Wales had the better of the first half-hour's play and it came as no surprise when they opened the scoring about nine minutes before half-time.

Dickie Owen gathered the ball at a scrum on the All Blacks' 25 and feinted to give an orthodox scrum-half pass to his partner Percy Bush. The New Zealand defence were anxious to bottle up the Welsh will-o'-the-wisp and, sensing he was to gain possession, swung towards the open-side anticipating an attack. Seeing that the defence was committed to Bush, Owen threw an accurate reverse pass to the narrow side where Cliff

Pritchard and Rhys Gabe completed a well-rehearsed move to work an overlap for Teddy Morgan to race along the left wing and score. Bert Winfield failed to convert, but Wales went into the interval nursing a slender 3–0 lead.

The tale of the second half is easily told. Time and again New Zealand attacked; time and again Winfield repulsed them with raking punts to touch. Near the end, however, came the incident that overshadowed the match and spawned a controversy that lasts to this day. From a run out of defence, New Zealand's Billy Wallace made more than 30 yards before passing to Bob Deans, who launched himself for the line. Many thought Deans had crossed, but the referee ruled he had been held up in the tackle, rejected claims for a score and shortly after whistled for time.

18

Longest winning streak

During the so-called first Golden Era of Welsh rugby between 1899 and 1913, Wales never lost at home against the other home Unions and, among all-comers, only South Africa (in 1906 and 1912) managed to beat the dragon on Welsh soil. The period also included Wales's longest peacetime span of successive victories (home or away) lasting nearly three years between March 1907 and January 1910:

Won 29–0	v Ireland 1907 (h)
Won 28–18	v England 1908 (a)
Won 6–5	v Scotland 1908 (h)
Won 36–4	v France 1908 (h)
Won 11–5	v Ireland 1908 (a)
Won 9–6	v Australia 1908 (h)
Won 8–0	v England 1909 (h)
Won 5–3	v Scotland 1909 (a)
Won 47–5	v France 1909 (a)
Won 18–5	v Ireland 1909 (h)
Won 49–14	v France 1910 (h)

Matches today take place more frequently and, although this record remains the longest measured in peacetime, 119 years later Warren Gatland's history boys set a new mark for the number of successive wins when they defeated England 21–13 in Cardiff to register their 12th on the trot. The winning sequence reached 14 (but in the considerably shorter interval of one year and five days) with the victory over Ireland at Cardiff that clinched the Grand Slam later in the season.

19

The Grand Slam

The Grand Slam of beating all opponents in the Five (and later Six) Nations tournament in the same season is the competition's most highly-prized achievement. Although the expression was not applied in a rugby context until 1957, Wales were the first nation to make a clean sweep of the other European unions when they defeated England, Scotland, France and Ireland in 1908. They have now won the accolade 12 times: 1908, 1909, 1911, 1950, 1952, 1971, 1976, 1978, 2005, 2008, 2012 and 2019.

20

BLOOD ON THE DRAGON

The First World War

Soon after war was declared on Germany in 1914, the WFU urged its clubs to appeal to the "pluck and patriotism" of its members and rally them to answer Lord Kitchener's call to enlist and serve King and Country. The response from clubs was immediate. Before the war was 30 days old the Cardiff club organised a meeting calling sportsmen in the area to get fit with a view to forming a sportsmen's battalion. At Newport, members of the rugby club were quick to swell the numbers of a local sporting platoon which became part of the South Wales Borderers, and many clubs surrendered their grounds to the military for training purposes or, in some cases, for conversion to allotments to sustain the war effort.

Among former Welsh international players, Billy Geen volunteered "as swiftly into the Army as he used to zig-zag through most defences". The fair-haired favourite from Newport was also among the first Welsh rugby nternationals who were casualties of the war that did not end all wars, but did bring down the curtain on international rugby for five years. The fallen were spread through every generation which had represented Wales on the rugby field, from Richard Garnons

Williams of the original 1881 Wales fifteen, through Charlie Pritchard of the famous 1905 pack against New Zealand and right up to Dai Watts of the immediate pre-war "Terrible Eight" – not forgetting those whose deaths from wounds sustained on active service fell outside the cut-off date designated by the Commonwealth War Graves Commission for deaths attributable to war service [31st August, 1921]. Men like Hop Maddock, the London Welsh speedster who succeeded Willie Llewellyn in the Welsh team of the Golden Era. Maddock was severely wounded serving with the Machine Gun Corps and died in Cardiff barely three years after the Armistice. Or Dai "Tarw" Jones, the magnificent Treherbert bull of the 1905 pack who joined up with the Welsh Guards and served on the Somme until a gunshot wound through his lung left him physically disabled. Jones lingered on bravely until meeting his untimely death, aged 51, on the same day that Wales finally laid the Twickenham "bogey" in January 1933.

Some organised rugby was played during the early years of the war, despite the suspension of official international matches. The biggest wartime encounter took place in April 1915 when Wales were beaten 26–10 by the Barbarians in a non-cap game arranged to recruit volunteers for the Welsh Guards and which raised more than £200 for military charities. The Welsh fifteen contained 13 pre-war caps and was again led by Rev. Alban Davies, now an Army chaplain, who was reunited with five of his "Terrible Eight" from 1914. Staged at Cardiff Arms Park and originally advertised as a Forces International between Wales and England, it was the only match played by a team that was representative of the Principality during the war-torn seasons from 1914 to 1918.

ROLL OF HONOUR 1914–1919

TAYLOR, Charles Gerald (Wales 1884–1887) Killed in action serving with the Battle Cruiser Squadron on HMS *Tiger* during the Battle of Dogger Bank on January 24, 1915.

GEEN, William Purdon (Wales 1912–1913) Killed in action serving with the King's Royal Rifle Corps at Hooge during the Second Battle of Ypres on July 31, 1915.

WILLIAMS, Richard Davies Garnons (Wales 1881) Killed in action serving with the Royal Fusiliers at the Battle of Loos, France, on September 25, 1915.

PHILLIPS, Louis Augustus (Wales 1900–1901) Killed in action serving with the Royal Fusiliers at Cambrai on March 14, 1916.

THOMAS, Edward John "Dick" (Wales 1906–1909) Killed in action serving with the Royal Fusiliers at Mametz Wood during the Battle of the Somme on July 7, 1916.

WILLIAMS, John Lewis (Wales 1906–1911) Killed in action serving with the Welsh Regiment at Mametz Wood during the Battle of the Somme on July 12, 1916.

WATTS, David (Wales 1914) Killed in action serving with the King's Shropshire Light Infantry at Ancre Valley during the Battle of the Somme on July 14, 1916.

PRITCHARD, Charles Meyrick (Wales 1904–1910)

Wounded while serving with the South Wales Borderers at the Battle of the Somme. Died in France on August 14, 1916.

THOMAS, Horace Wyndham (Wales 1912–1913) Killed in action serving with the Rifle Brigade at Guillemont, France, on September 3, 1916.

LEWIS, Brinley Richard (Wales 1912–1913) Killed in action serving with the Royal Field Artillery at Ypres on April 2, 1917.

WESTACOTT, David (Wales 1906) Killed in action serving with the Gloucestershire Regiment at Zonnebeke in Flanders, Belgium, during the Third Battle of Ypres on August 28, 1917.

WALLER, Phillip Dudley (Wales 1908–1910) Killed in action serving with the South African Heavy Artillery at Arras, France, on December 14, 1917.

PERRETT, Fred Leonard (Wales 1912–1913) Wounded while serving with the Royal Welsh Fusiliers. Died in France on December 1, 1918.

21

FAMOUS FIRSTS

Through the card

Wales's first Five Nations match after the Great War was at home to England at Swansea in January 1920 when they scored an unexpected win by 19–5 in difficult conditions. It was a day to remember for Jerry Shea, the Newport centre creating a new record for international rugby by becoming the first player to go through the card of scoring actions in a Test. His haul of 16 points included a try, conversion, penalty goal and two dropped goals.

Even so, he received mixed reviews in the aftermath, including harsh criticism that his play was selfish. He was dropped from the side after their next match and shortly before Christmas 1921 boarded the bandwagon that carried so many of his promising Welsh contemporaries to the rugby league lands of the north, Shea throwing in his lot with Wigan.

STAT ZONE

Wales's oldest player

When Newport's scrum-half, Tommy Vile, captained Wales against Scotland at Swansea in February 1921 he became, at 38 years and 149 days, the oldest player to represent Wales. His Welsh career stretched over 13 years, for he had won his spurs as a youngster against England at a fog-bound Bristol in 1908, but as a Lion he had played three Tests in Australia and New Zealand with the 1904 British/Irish team. All told, then, his Test career spanned more than 16 years, a staggering record in those amateur days.

STAT ZONE

Three times for a Welshman

The rarest score in rugby was the goal from a mark, a goal kicked after a fair catch was claimed in open play. In the early days the kick at goal could be taken by any member of the team, not solely the player making the mark (as was the case in later years). The scoring action ceased to exist after 1977 when the free-kick clause was introduced and a mark could only be awarded inside a defending player's 22-yard area. The only three goals from marks were kicked for Wales in international matches:

Player	Opponents	Venue	Year
WJ Bancroft	Scotland	Inverleith	1895
HB Winfield	England	Leicester	1904
WC Powell	England	Twickenham	1931

24

STAT ZONE

Unchanged through the season

Wales have been involved in the International Championship (initially involving four, then five and now six nations) since the tournament began unofficially in the winter of 1882/83. Twice the fifteen has been unchanged for all of its matches: in 1932 and 1972.

France were expelled from the Five Nations after 1931 so Wales played only three Championship matches in 1932. The team comprising full back and captain Jack Bassett; threequarters Jack Morley, Claud Davey, Frank Williams and Ronnie Boon; Raymond Ralph and "Wick" Powell at halfback; and forwards Archie Skym, "Lonzo" Bowdler, Tom Day, Dai Thomas, "Ned" Jenkins, Will Davies, Watcyn Thomas and Arthur Lemon was retained en bloc for the victories against England and Scotland and the late defeat by Ireland that robbed them of the Triple Crown.

The team of all the talents that went unaltered and unbeaten through the 1972 Championship was JPR Williams; Gerald Davies, Roy Bergiers, Arthur Lewis, John Bevan; Barry John, Gareth Edwards; John Lloyd (captain), Jeff Young, Barry Llewelyn, Delme Thomas, Geoff Evans, Dai Morris, John

Taylor and Mervyn Davies. They beat England, Scotland and France comfortably but political strife in Ireland prevented the WRU from fulfilling their fixture in Dublin. The same Welsh fifteen was selected for a match scheduled for 11th March that year, but to the disappointment of both sides the game was postponed and the Championship was never resolved.

25

FAMOUS CHARACTERS

The sacrificial car of Juggernaut

The giant of Welsh rugby in the thirties was the tall, long-striding threequarter, Wilf Wooller. A native of North Wales, where he was educated at Rydal School, he made such an impact on the selectors whilst playing for Sale that they drafted him into the international side as a teenager.

The 6' 2" Wooller proved the selectors' faith in him, playing a strong defensive role in the Welsh side that, in 1933 and at the tenth attempt, finally managed its first win at Twickenham. But it was when he joined Cardiff during his Cambridge days that the best was seen of him, playing alongside the will-o'-the-wisp fly half, Cliff Jones. Wooller could win matches with wonderfully creative play, often scoring near-impossible tries or landing towering goals. Yet he could also save apparently certain tries by delivering the most lethal of crash tackles. He was still in his prime when the outbreak of war in 1939 effectively brought his international rugby career to an end.

Even so, it was his sheer determination and never-say-die attitude that helped Wales to its most famous victory of his era, against New Zealand in 1935. With Wales trailing 12–10 and a man short, Wooller's opportunism led to the thrilling late winning try. He made the carefully-judged kick ahead that deceived the All Blacks and allowed Geoffrey Rees-Jones in for the decisive score.

The *Daily Telegraph*'s Howard Marshall, the leading critic of the day, described Wooller's talents as "like the sacrificial car of Juggernaut, leaving a trail of prostrate figures in his wake."

26

Welsh rugby's only Test cricketer

Although Wilf Wooller went on to distinguish himself as a cricketer, helping Glamorgan to their maiden County Championship title in 1948, it was another Welshman in that first side to storm the Twickenham citadel who holds the unique distinction of turning out at rugby for Wales and playing Test cricket. Maurice Turnbull, the Cardiff scrum half who partnered Harry Bowcott in that 7–3 Twickenham triumph in 1933, appeared in nine cricket Test matches for England between 1930 and 1936. Both of his Welsh rugby caps were won in 1933.

STAT ZONE

An expensive headwear bill

There were obviously 15 new caps in the Welsh side that was the first to take the field in 1881, and one can understand why as many as 13 players made their international debuts in 1919 and 1947 when Wales staged their first official post-war Tests after five-year and eight-year breaks caused by the respective World Wars. But arguably the most unexpected headwear bill the WRU ever forked out was in January 1934 when their selectors, in a rare moment of rashness, awarded 13 new caps for the encounter with England. The new blood was introduced barely ten months since their previous international engagement, but the experiment proved a distinct failure. Not only were Wales beaten 9–0 at Cardiff, five of those 13 debutants were never again called on to represent their country.

FAMOUS FIRSTS

The full back who caused a stir

The match against Ireland at Swansea in March 1934 was the 163rd international played by Wales since 1881. It was also the first Welsh match in which their full back crossed for a Test try. The game appeared to be heading for a rare scoreless draw until full back Vivian Jenkins joined the Welsh threequarter line to initiate a handling move which culminated in him scoring a try in the corner which he himself converted from near the touchline. A flurry of points followed for Wales to finish winners by 13–0, but Jenkins's action caused quite a stir. The duties of full backs, it was argued, were primarily defensive: catch the high ball, fall on the rolling ball, kick for touch and tackle would-be try-scorers. The next full back to score an international try for Wales was Keith Jarrett against England in 1967 before that famous number 15, JPR Williams, made a habit of scoring tries after the law restricting direct kicking to touch was introduced in 1968.

FAMOUS CHARACTERS

The schoolboy who beat the All Blacks

"Don't tell them back home that we were beaten because of schoolboys," New Zealand skipper Jack Manchester is reputed to have said after Swansea had defeated his Third All Blacks in September 1935. The schoolboys were the Gowerton Grammar School cousins Haydn Tanner and Willie Davies, Swansea's half backs. Within two months of that match, Tanner was back orchestrating another defeat of the All Blacks, this time on his debut for Wales.

At scrum half he made a positive impact in the Welsh side with his quick service, strong defence and infallible option-taking. He was never dropped in a long career that spanned 14 seasons and made valuable contributions to three Welsh sides that either shared or won the International Championship outright.

But for the war, Tanner would certainly have set an incredible cap record for his country. He returned in 1947 to lead Wales a dozen times, missing only the match against Australia in 1947 when he was forced to withdraw through an arm injury. As captain he was a shrewd tactician who dictated the course of the game and inspired average players to perform beyond their potential.

Writing in 1954 the leading Welsh critic JBG Thomas called the tall, powerful 14-stoner the greatest scrum half of them all. "His immaculate touch; his unerring accuracy and his unobtrusiveness demand for him a place among the immortals," he wrote.

FAMOUS GAMES

That lovely point

WALES 13, NEW ZEALAND 12
Cardiff Arms Park, 21 December 1935

The outstanding Welsh performance between the wars came in their match against the Third All Blacks. Wales had been well beaten at Swansea by the 1924 tourists and this match back at Cardiff, where Wales had beaten the 1905 Originals, was viewed by the nations as the series decider.

Wales fielded a side that was a mixture of youth and experience, taking four new caps into what turned out to be a thriller. Wales led 10–3 early in the second half before a spirited recovery took the New Zealanders 12–10 ahead with time running out. Then the huge crowd was stunned into silence when Swansea hooker Don Tarr was stretchered off with a broken neck, leaving Wales with a steep uphill task in front of them.

This Welsh side, however, lacked nothing for character. Wales pressed and Wilf Wooller put in a well-judged high punt that landed over the All Blacks' line. The ball bounced eccentrically and with Wooller having overrun it, many felt for a split second that the chance of victory was lost. But following up behind was Geoffrey Rees-Jones, the right wing, and he scored the try that gave Wales victory by what pack-leader Arthur Rees for years after described as "that lovely point".

This side went on to enjoy its best Championship season of the thirties, winning the title outright for the first time for five years.

31

The last four-point dropped goal

The Principality's last international match before World War Two was a 7–0 win against Ireland in Belfast that secured Wales a three-way share of the 1939 International Championship. All the Welsh points were scored in the last five minutes by their fly half, Willie Davies. He went over for his first international try, four minutes after dropping the first goal of his senior career. The goal was the last four-pointer by a Welshman in a Test. The value of the scoring action was reduced to three points in 1947.

BLOOD ON THE DRAGON

The Second World War

The 1939–40 rugby season in Wales should have been the most intense International campaign for 12 years. France, separated rather than divorced from the Five Nations since 1931 on account of misgivings held among the Home Unions about the extent of professional practices on the continent, had finally been invited to rejoin the International Championship, and a full-scale tour of Europe by the Wallabies was arranged. However, the season turned out to be the shortest on record. After just one first-class match in Wales, between Cardiff and Bridgend played on Saturday 2nd September 1939, Britain declared war on Germany the next day, and for the second time in 25 years the WRU suspended organised rugby.

Mercifully fewer were killed in action between 1939 and 1945, but the three Welsh internationals who were casualties included two who had made their sole appearances among the 13 new caps at Cardiff in 1934: Cecil Davies and John Evans. Evans was a former Junior Schools and Secondary Schools cap and had skippered the Welsh team on his only Test appearance. The third former Welsh cap who fell in the war was Maurice Turnbull, who had played scrum half in the first Welsh victory at Twickenham in 1933.

His famous Glamorgan cricketing colleague Wilf Wooller (who led Cardiff and Wales XVs in the first wartime rugby season) became Welsh sport's best-known prisoner-of-war. Wooller was commissioned into the Royal Artillery before going out to Java early in 1942. There was no news of his whereabouts for several months and he was officially posted as missing until reports reached Wales that he was being held by the Japanese. When he returned home nearly four years later he was a mere shadow of the strapping threequarter who owed his rugby fame to those prodigious deeds against the 1935 All Blacks. But while he never played rugby seriously again, happily in 1948 his recovery was so complete that he led Glamorgan to an historic first County Championship cricket title.

ROLL OF HONOUR 1939–1945

DAVIES, Cecil Rhys (Wales 1934) Killed in action during operations with the Royal Air Force over Brittany, France, on December 24, 1941.

EVANS, John Raymond (Wales 1934) Killed in action serving with the Parachute Regiment near Tunis in North Africa on March 8, 1943.

TURNBULL, Maurice Joseph Lawson (Wales 1933) Killed in action with the Welsh Guards near Montchamp, Normandy, France on August 5, 1944.

WARTIME RUGBY

The Second World War

The significant social lesson grasped during World War One had been the important role sport played in maintaining civilian morale in times of gloom and hardship. This was not forgotten during the second conflict and sport, through the organising abilities of the services, continued more extensively than in 1914. To all intents and purposes rugby in Wales bore comparison to the activities of happier pre-war days. By the end of September 1939, an informal season was announced and many fixtures took place between service units, students and even the clubs. Cardiff, in particular, made huge efforts to continue playing, actively recruiting from reserved occupations, including mining, and servicemen stationed nearby. The policy paid off. Emerging talents of the calibre of Bleddyn Williams, Jack Matthews, Maldwyn James, Frank Trott and Billy Cleaver were introduced to senior rugby. All skilful players, they would progress confidently to provide ready-made experience to both the Arms Park club and Wales when normal service was resumed in 1945. Another platform for promising players was provided through the initiative of C.B. Jones of the Swansea *South Wales Evening*

Post newspaper. He raised numerous South Wales service fifteens of international quality to keep the rugby flag flying in the west through matches staged on the St Helen's ground.

Once again a relaxation of professional bylaws permitted rugby league players serving in the forces to play the union game. Between two and three dozen Welsh Internationals capped in the thirties had gone north and many of them were welcomed back with open arms during the period of hostilities. Former internationals Willie Davies, Idwal Davies and Eddie Watkins were among the most recent converts who featured in many representative union matches together with star Welsh RL exports who included the famous Jim Sullivan and Gus Risman.

Red Cross Internationals between England and Wales were organised in 1939/40 before giving way to a long series of Services Internationals in the middle years of the war. Then, when peace was declared, a season of "Victory Internationals" followed in 1945/46 when rugby in Britain and Ireland was given a tremendous shot in the arm by the visit of a New Zealand Forces side known as the Kiwis. Drawn from New Zealand rugby players who had been stationed in North Africa and Europe during the war, the side had the cache and playing uniform of full-blown All Blacks, and generated a distinguished record to match. They drew 3–3 with Newport and were beaten only three times in 38 games in Europe – once by Monmouthshire (15–0) – but pride of place among their wins was an 11–3 success against a Wales XV at Cardiff Arms Park, the scene of the defeats suffered by their All Black predecessors of 1905 and 1935.

None of the "International" matches played between 1940 and 1946 warranted full caps because, it was argued, the best players weren't always available owing to national service calls. Even during the post-war "Victory" series many were still stationed abroad awaiting demob. Take the case of Haydn Tanner. He was the automatic choice to lead Wales but was often unable to obtain sufficient leave to travel from Austria to play in Britain. Just imagine what a magnificent Test appearance record he might have had had war not interrupted his career.

Wartime and Victory International results:

Red Cross Internationals (1939–40)
9 Mar 1940   Lost 9–18 v England (Cardiff)
13 Apr 1940   Lost 3–17 v England (Gloucester)

Services Internationals (1942–45)
7 Mar 1942   Won 17–12 v England (Swansea)
28 Mar 1942   Won 9–3 v England (Gloucester)
7 Nov 1942   Won 11–7 v England (Swansea)
20 Mar 1943   Won 34–7 v England (Gloucester)
20 Nov 1943   Won 11–9 v England (Swansea)
8 Apr 1944   Lost 8–20 v England (Gloucester)
25 Nov 1944   Won 28–11 v England (Swansea)
7 Apr 1945   Won 24–9 v England (Gloucester)

Victory Internationals (1945–46)
22 Dec 1945   Won 8–0 v France (Swansea)
5 Jan 1946   Lost 3–11 v NZ Kiwis (Cardiff)
19 Jan 1946   Lost 13–25 v England (Cardiff)
2 Feb 1946   Lost 6–25 v Scotland (Swansea)
23 Feb 1946   Won 3–0 v England (Twickenham)

9 Mar 1946 Won 6–4 v Ireland (Cardiff)
30 Mar 1946 Lost 11–13 v Scotland (Murrayfield)
22 Apr 1946 Lost 0–12 v France (Paris)

BEFORE AND AFTER

Capped either side of the war

Four Welsh players' cap careers spanned the long break caused by the Second World War:

Player	Career span
Haydn Tanner	(1935–1949)
Bill Travers	(1937–1949)
Howard Davies	(1939–1947)
Les Manfield	(1939–1948)

FAMOUS CHARACTERS

The fastest man in Wales

Ken Jones was the fastest Welsh wing since the golden days of Willie Llewellyn and Teddy Morgan, weaving a web of magic in a ten-year Test career during which he set a national appearance record by winning 44 caps. The secret of the prolonged success that brought him the 17 Test tries that matched the then national record set before World War One was his sense of balance. He was the perfect study in concentration as a runner and, at a time when defences engaged in tight man-to-man marking, he had the initiative to create his own chances.

He was at his peak in the Grand Slam years of the early fifties. He scored four tries in 1950 when Wales claimed the prize for the first time for 39 years, and the four he collected in 1952 included two that demonstrated his genuine pace and instinctive talent. At Twickenham he streamed past two defenders for a classic score. Then, in the Triple Crown match in Dublin, he showed his wit by moving in off his wing to take a short pass from Cliff Morgan and make a thrilling 50-yard run to the try-line. Arguably, though, his match-winner against New Zealand in 1953 attracted the largest audience. In the first Cardiff Test broadcast live to a Welsh audience, Jones showed the instinctive genius of a player who knew the right place to be at the right moment. Latching on to Clem Thomas's cross-kick, he raced around for the try that gave Wales their last win to date against the All Blacks.

When, in January 1957, Ken Jones was dropped from the Welsh right-wing berth for the first time since the war, he had set a remarkable record of 43 consecutive Tests for the Principality – ten more than the previous record-holder, Billy Bancroft, who was an ever-present between 1890 and 1901. To date, only Gareth Edwards has played more successive cap matches for Wales. The players who have made the most appearances without a break for Wales are:

Player	Run of caps	Relevant span
Gareth Edwards	53	1967–1978
Ken Jones	43	1947–1956
Graham Price	39	1975–1983
Mervyn Davies	38	1969–1976

37

FAMOUS CHARACTERS

The Cardiff legend

The player who dominated British midfield play in the immediate post-war seasons was the Cardiff and Wales rugby legend, Bleddyn Williams. Watching him play was like attending a master-class. He perfected every basic skill and was a shrewd reader of the game, best-remembered for his effective side-step. Williams made his Wales debut as a fly half in their first official post-war international, against England in 1947. Thereafter he became a regular choice in the centre, the position where he helped Cardiff to become the dominant club force in British rugby in the late 1940s.

In 1950 he was given the honour of succeeding Haydn Tanner as captain of Wales for the opening match of the Championship at Twickenham. But less than 24 hours before kick-off, he had to withdraw through injury. Wales were captained instead by John Gwilliam and such a success he made of the job that Williams had to wait until two Grand Slams and three years passed before he was again offered the honour. Indeed, Williams only played three times for Wales between 1950 and the start of 1953. He eventually took over from Gwilliam as captain after the defeat by England in 1953 and proceeded to take Wales through a wonderful run of five wins in five matches, including the 13–8 defeat of New Zealand at Cardiff in December 1953.

38

The changing value of the dropped goal

For more than 75 years the dropped goal was the king of rugby's scoring actions. Before points were introduced, a majority of goals determined results, so a single dropped goal was worth more than any number of unconverted tries. Then, when points were introduced to decide international matches in the late 1880s, the dropped goal carried a higher value than a try. It was not until 1948 that the International Board reduced the value of the scoring action from four points to three, putting the drop on a par with a try. Wales's first three-point dropped goal was scored by the Cardiff fly half Billy Cleaver, against Scotland at Swansea in the Triple Crown season of 1950.

FAMOUS CHARACTERS

The lineout expert

The most skilful and athletic Welsh jumper of the twentieth century was Neath's Roy John, who expertly exploited the line-out laws of his day. He came into the side as a replacement for his club colleague Rees Stephens against England in 1950 and lorded it over the best in the world for four years. His spring-heeled jumping made him one of rugby's great entertainers, though that's not to underestimate the asset John was to the Grand Slam teams of 1950 and 1952. He ensured Wales took complete control of the throw-ins and the wealth of possession he brought enabled his backs to exercise a tight tactical grip on matches.

Yet he was not just a brilliant line-out man. He had good positional sense, often turning up unexpectedly in the loose to provide an essential link. He also possessed pace and could execute all the skills of the game. In Dublin in 1952, for example, it was his perfect dummy and exquisite side-step that paved the way for Clem Thomas to score the try that sealed Wales's Triple Crown win. A year later he was a member of the side that overcame the All Blacks at Cardiff, but by the end of the 1954 Five Nations he had given way to Rhys Williams. The fact that the International Board changed the line-out rules in 1954 to outlaw the levering that had been such an important aspect of his technique was seen in Wales as a deliberate move to check their line-out supremacy. There could be no better reflection of Roy John's skills.

FAMOUS GAMES

The first Triple Crown for 39 years

IRELAND 3, WALES 6

Ravenhill Park, Belfast, 11 March 1950

Before France became a truly recognisable force in the Five Nations at the end of the fifties, the Holy Grail of the annual International Championship was winning the Triple Crown – beating all of the other Home Unions. Wales's last win of the then mythical trophy had been in 1911 when they met Ireland in Belfast in 1950 with the famous trophy within their grasp. Ireland, who had won the Triple Crown in 1948 and 1949, engaged Wales in an absorbing power struggle between two packs that had similar virtues. Ireland's successes in the previous seasons had been based on the subdue-then-penetrate theory that was best summed up in the vogue phrase "Triple Crown Rugby". Now it was Wales under the intelligent leadership of No 8 John Gwilliam who sought to beat Ireland at their own game in order to win the Crown.

A protracted forward battle yielded no scores for nearly an hour. Then the Welsh back row induced an error by the Irish halves, Jack Matthews picked up the ball and launched Ken

Jones on a 20-yard foot race to the corner for a try. Wales led for only three minutes, George Norton kicking three points when the forwards were penalised at a maul. With time running out Malcolm Thomas broke the stalemate when he crashed through a tackle to squeeze in at the left corner. In the mayhem the referee had been unsighted, but the Irish touch-judge confirmed the try was valid and Wales celebrated winning the Triple Crown for the first time for 39 years.

41

The unique Mr Gwilliam

A fortnight after securing the 1950 Triple Crown, Wales went on to overwhelm France 21–0 at Cardiff and win the Grand Slam for only the fourth time. Their clever skipper was the schoolmasterly John Gwilliam, who had served as a lieutenant with the Royal Tank Regiment in the after-waves of the 1944 Normandy landings that set in train the liberation of Western Europe. After demob, he returned to Cambridge to complete his education, won Blues in 1947 and 1948, and became a history teacher. Austere and aloof almost to the point of being remote from his colleagues, he nevertheless commanded the respect of all who came into professional or playing contact with him. The strength of his personality and the magnificent team spirit he engendered paid handsome dividends for Welsh rugby. He had not been the selectors' first choice as captain but he grasped the chance to lead Wales to the Triple Crown and Grand Slam with open arms after Bleddyn Williams and Rees Stephens, the leading candidates for the job, withdrew from the fifteen originally selected to meet England at Twickenham at the start of the campaign. Gwilliam's unique style of leadership was captaincy by intent. Rather than issue strict instructions, he set the parameters within which his side was to operate and ensured that team members knew how to contribute to the greater good of the fifteen. He urged them to take responsibility for their actions and by doing so empowered them to make decisions in the heat of the moment. Gwilliam's approach gave players confidence and he retained the captaincy two years later when his side again brought home the Grand Slam, making him the only man to lead Wales throughout two Grand Slams.

42

FAMOUS CHARACTERS

The King of Rugger

Lewis Jones had a meteoric rise to fame as a teenager. Within six months of making his first-class club debut he was in the Welsh team that defeated England at Twickenham in 1950. He played like a veteran, scoring a penalty and converting a try, and went on to become the youngest Welshman to date to feature throughout a Grand Slam campaign. He was still in his teens when he made an outstanding contribution to the Lions tour of New Zealand and Australia later that year.

Jones had all the skills. He was an incisive runner, a prodigious kicker and was the model of steadiness in defence. The Welsh selectors could not afford to omit the youngster from the side and took the then very unusual measure of choosing him as a full back, centre and wing. It was, they believed, imperative to accommodate him somewhere in the back division. Jones repaid their faith in him by amassing 36 points in only ten games, a creditable return given the low scoring that was a feature of international matches in the fifties. Wales lost only twice when Jones was in their ranks: the extraordinary 0–19 defeat by Scotland in 1951 and the narrow 3–6 slip against the Fourth Springboks later the same year. After contributing significantly to the 1952 Grand Slam, he took his skills north to Leeds. When he finally retired from playing, he could truly call himself the King of Rugger, the title of his subsequent autobiography, for he became an equally distinguished exponent of the league code.

43

FAMOUS CHARACTERS

The great tactical controller

The automatic choice as Wales fly half in the 1950s was the Rhondda's Cliff Morgan. He played 29 times to set a record (that stood for nearly 40 years) as his country's most-capped player in the position. Early in his career his kicking was a weakness – and it is a remarkable fact for a Test fly half that he never dropped a goal in his entire international career – but he more than compensated for that with his quick-wittedness as a runner. He was a shrewd tactician who showed his ability as an option-taker as early as his second season in the Welsh side. Morgan's tactical control of the Welsh midfield in 1952 was a telling factor in the Grand Slam triumph and his sharp change of pace when sensing openings set up important tries in the tough away wins against both England and Ireland that year.

A huge box-office attraction throughout his career, Morgan steered Wales to three more Championship titles (1954 shared, 1955 shared and 1956 outright when he was captain) and was the spark that ignited the Cardiff and Wales back divisions to memorable victories over the Fourth All Blacks in 1953. He also flourished on the firm South African grounds with the 1955 Lions, his electric displays sending spectators into raptures. To the astonishment of his followers, however, he retired aged only 27 in 1958. He sensed that some of his sting had evaporated and chose to quit while in his prime. Morgan never regretted the decision.

44

FAMOUS GAMES

Still unbeaten at Cardiff against New Zealand

WALES 13, NEW ZEALAND 8

Cardiff Arms Park, 19 December 1953

In one of the first Cardiff internationals televised live by the BBC, Wales staged a storming finish to retain their then unbeaten Cardiff record (three wins in three games) against the All Blacks. The tourists looked odds-on favourites to win for most of the match. On top up front, the New Zealanders led 8–5 at half-time and were comfortably in the saddle as the game entered its final quarter. To compound Wales's problems, their centre Gareth Griffiths was off the field receiving treatment for a dislocated shoulder.

His second-half return, however, heralded a magnificent Welsh revival. Gwyn Rowlands levelled the scores with a penalty goal from in front of the posts and, nine minutes from time, Griffiths figured in the crucial move of the match. He put pressure on the New Zealand defence inside their 25 and when the ball came back on the Welsh side, Clem Thomas put in an inspired cross-kick. Ken Jones, far out on the right, latched onto the bounce and swerved past Ron Jarden for the winning try. Wales haven't beaten the All Blacks since.

45

FAMOUS CHARACTERS

Good enough to be an honorary All Black

When the International Board outlawed the so-called wedge method of patterned blocking at the line-out in 1954, a new breed of jumper able to hold his own in the absence of protectors was required. The player who admirably filled that role for Wales was beefy Rhys Williams. In the late fifties, line-outs often developed into protracted standing mauls. There, Williams's upper-body strength was priceless. A science graduate, he brought a systematic and intellectual approach to his work in the second-row, giving regular master-classes in line-out catching, scrummaging and even the ancient rugby art of dribbling.

Wales prospered while he worked in the engine room. They shared the Championship title in each of his first two seasons in the side and were outright champions in 1956, when Williams's athleticism won plenty of possession in the tight and his aggressive defence held up opponents who threatened danger in the loose. After that, the widely-held tenet among the critics was that the best of big Rhys was never seen in the Five Nations. He enjoyed a hugely successful tour of South Africa with the 1955 Lions, and the magnificent New Zealand pack against whom he toiled tirelessly for the 1959 tourists greeted him as a 'blood-brother' – conferring honorary All Black status on him after his exploits in a tough four-Test series against the likes of Colin Meads, Wilson Whineray and Kel Tremain. But after leading Wales to a disappointing defeat at Twickenham in 1960, he was dropped and soon after retired as his country's then most-capped lock forward.

46

STAT ZONE

The most successful Welsh international of all time?

Many Welsh players finished their careers with 100% winning Test records, but the only one who did so by scoring the winning points in every game he played in was the Llanelli centre Denzil Thomas. Admittedly he only played once, but it was an unforgettable day for the little man they called "Denzil Drop". His hour of glory came in March 1954 when, with Wales holding Ireland 9–all in a try-less game in Dublin, Thomas broke the deadlock by dropping a late goal that gave his side a 12–9 win. But alas poor Denzil: he was dropped for Wales's two remaining matches of the season, against France and Scotland, and never came close to winning international honours again.

47

FAMOUS CHARACTERS

The fastest striker in Wales

Bryn Meredith was the fast-striking hooker whose scrum possession helped Cliff Morgan direct Wales to Championship honours between 1954 and 1956. Early in his career he formed a formidable front-row with Billy Williams and Courtenay Meredith (no relation) for both Wales and the Lions, the three developing such an excellent scrummaging technique that it was extremely rare to see them concede a scrum against the head in the 13 major internationals they played together. Bryn was athletic in the line-out (long before the days when hookers became throwers-in) and a mobile operator with safe hands in the loose. His perceptive sense of anticipation in broken play brought him three Test tries, a respectable return from an era of tight marking and low scoring at international level.

Honours eluded Wales after 1956, but Bryn Meredith had the fitness and stamina to remain in prime condition for eight years at Test level. Given the physical punishment hookers endure, his overall record makes impressive reading. He captained Wales four times and all told was the cornerstone of the pack in 34 matches to finish his career as Wales's then most-capped forward.

48

Entire Welsh XV sent off

When Wales played Ireland in appalling conditions at Cardiff Arms Park in March 1957, the Scottish referee Jack Taylor became so frustrated at being unable to identify the teams from their mud-bespattered jerseys that he ordered the entire Welsh side from the field … but only to change their kit. Wales, it seems, benefited more from the unexpected change. Trailing 5–3 at the time, they returned to dig out a 6–5 victory thanks to a late penalty kicked by the Llanelli full back, Terry Davies.

49

FAMOUS CHARACTERS

A firm favourite

A reputation built on the college rugby circuit propelled Dewi Bebb to the top of the selectors' lists in 1958–59. He enjoyed a successful final trial and was immediately promoted to the national side for the opening match of the Five Nations Championship. Nothing endears a Welshman more to his people than to score a try against England. Bebb scored the winner on his debut, and went on to score ten more tries for Wales – including another five against England – in a distinguished career spanning nine seasons.

A firm favourite with crowds, he was a level-headed player and an outstanding specialist left wing. His dashing opportunism stamped him as the leading try scorer at a time when expansive threequarter play and easy run-ins for wings were at a premium. He was a permanent fixture in the 1964 side that shared the Championship title with Scotland and he scored the try that turned the Triple Crown showdown with Ireland at Cardiff in 1965. Early in the second half he snaffled up a deflection from a line-out on the Irish line to touch down in the corner.

Although dropped for the early part of the 1966 campaign, the selectors felt in dire need of his speed to take on the French at Cardiff in March. Bebb's pace on a firm pitch helped his nation to the Championship title and earned for him a place on the Lions tour to New Zealand. He signed off in 1967 where he had started: scoring a try against England in a wonderful open game at Cardiff. That game showed that had he played under more open laws then his haul of international tries would have easily doubled.

FAMOUS CHARACTERS

The modern back row forward

A blend of speed and intelligence distinguished Alun Pask as an international forward. True, critics felt that he shunned the donkey work at rucks and mauls, but his mobility and safe hands equipped him well for attacking roles. He was the first Welshman to appreciate the tactical potential of the ploy, the fast back row moves launched from scrums and line-outs so expertly demonstrated by the French between 1959 and 1962, when they ruled the Five Nations roost. Pask was ideally suited to slip into this high-gear style and perfected it as a method of attack. But there was more to his game. He could cover, he could tackle, and occasionally showed that he could sell a credible dummy. Once, during a critical early period in the Triple Crown match with Ireland in 1965, he even deputised as full back when an injury to John Dawes necessitated a rearrangement of the Welsh back division. Pask, under pressure, was the model of security in difficult conditions until Dawes returned. The first of the modern style of back rower to feature for Wales, all his successors based their game on Pask's.

He served a long apprenticeship as a Welsh travelling reserve, sitting out 13 matches before getting his chance in Paris in 1961.

He went on to win 26 successive caps as a versatile forward equally at home as flanker or No 8. He was a key member of the sides that shared the Championship in 1964 and won it outright in both 1965 and 1966, when he was captain. Unlucky not to be nominated captain of the 1966 Lions to New Zealand, he retired suddenly midway through the 1967 campaign.

51

Wales's lowest-scoring results

Wales's visit to Twickenham in 1962 resulted in the last of Wales's four scoreless draws. A game described as "stillborn" by the press was marred by slow heeling by the packs and excessive kicking from the backs, and the Welsh place-kicker missed five penalty attempts in the Principality's lowest-scoring result. Mercifully, Wales haven't been involved in a nil-nil stalemate since that drab January day. The only scoreless draws involving Wales were against Scotland in 1885, and against England in 1887, 1936 and 1962.

52

FAMOUS CHARACTERS

Top Cat

He wasn't the sharpest scrum half around the field and he never claimed to possess the fastest service, but for standing his ground in the face of marauding forwards Clive Rowlands had no equal. Even so, his selection as captain on debut in 1963 came as a surprise. In charge of an inexperienced team, he had to swallow his pride after losing to England. But his determination to turn the side into winners shone through a fortnight later when he dictated the match with Scotland from start to finish. Rowlands played to his pack at every opportunity and a match that yielded the staggering number of 111 line-outs ended 6–0 to Wales – their first Murrayfield victory for a decade.

Rowlands remained Wales's "Top Cat" for two more seasons, leading a team that remained largely unaltered with an iron fist. He managed to imbue a limited side with a corporate will to succeed and his methods quickly paid dividends. They shared the Championship in 1964 and, despite a humiliating defeat in South Africa, Wales returned to win the Triple Crown and Championship in 1965. Then, at the start of the following season he was dropped as surprisingly as he had been originally selected, his run of 14 matches as captain spanning a career that is unique in British rugby. Later he was the inspiring national coach of a side that dominated the Five Nations in the late sixties and early seventies.

53

STAT ZONE

Captains on debut

Clive Rowlands was one of only a handful of players who captained Wales on their international debuts.

Captain	Debut	Result
James Bevan	1881 v England	Lost
Charlie Lewis	1882 v Ireland	Won
John Evans	1934 v England	Lost
Clive Rowlands	1963 v England	Lost
Mike Watkins	1984 v Ireland	Won

To put the achievements of the last three into perspective, it is perhaps worth noting that Bevan and Lewis led in Wales's first and second international matches.

INFAMOUS GAME

The match of 111 line-outs

SCOTLAND 0, WALES 6
Murrayfield, Edinburgh, 2 February 1963

After a run of disappointing performances in the early sixties, nobody plotted more deviously to restore Wales's winning ways in 1963 than Clive Rowlands at Murrayfield, where single-handedly – or more accurately, single-footedly – he was responsible for a Test match that yielded the staggering number of 111 line-outs. It was not good theatre. In years to come he would be hailed as a national treasure, but as far as the media and many supporters were concerned, in February 1963 he was public enemy number one for his playing approach. Still, with Wales winning north of the Tweed for the first time for a decade, Rowlands felt the end justified the means.

Clive Rowlands: Wales travelled to Scotland and to Murrayfield [and] naturally, I was delighted that I'd been retained [as captain]. Prior to the appointment of a national coach, much responsibility rested on the captain's shoulders as he strove to unite individual strengths into collective combat.

I was fortunate in having Roy Bish at hand while also utilising the skills of fellow teachers Alun Pask and Brian Price. Alun was my vice-captain and pack-leader while I retained overall tactical responsibility. In Scotland I changed the tactics after ten minutes and it brought success. As captain of Wales I wanted Wales to win at all costs within the laws. I believed in playing fairly but hard.

FAMOUS FIRSTS

On tour

The summer of 1964 was notable for Wales's first international tour. Scotland, Ireland and England had pioneered short rugby visits to the southern hemisphere in the early sixties, but Wales's trip to South Africa in May 1964 was rendered more interesting by the IRB's recent introduction of sweeping new laws designed to encourage more open rugby. The most significant amendment related to the off-side line at the line-out which was moved to ten yards behind the line of touch for both back divisions. The change allowed for more enterprising back play, but Wales were unready to exploit it and a thumping 24–3 reverse – the worst Welsh performance for 40 years – set the Principality debating about the state of its national game. The defeat which had exposed Wales's lack of fitness and tactical shortcomings turned out to be a blessing in disguise because the urgent review that was immediately ordered led to a coaching revolution with Ray Williams appointed as national organiser. Rugby in Wales was mapping out a new course.

56

Winner takes it all

WALES 14, IRELAND 8

Cardiff Arms Park, 13 March 1965

Wales and Ireland were both playing for the Triple Crown, the first winner-takes-all showdown involving the countries since 1911. Sadly the weather detracted from the occasion, a steady downpour dampening the proceedings of a match that beforehand had generated huge interest.

Wales had their backs up against the wall in the first half. Ireland missed three easy chances of penalty points and failed to take advantage when John Dawes was off the field injured for 20 minutes. During his absence Alun Pask withdrew from the pack and proved a very accomplished performer at full back in a re-arranged back division. Wales lifted the siege just before half-time when David Watkins's long diagonal kick set up his namesake, Stuart, for the opening try. Terry Price, the teenaged Llanelli full back, converted. In the second half Price also dropped the goal from near halfway that put Wales into a safe 11–3 lead. "A good one, a very good one, and it's over!" G V Wynne-Jones told the huge radio audience.

Clive Rowlands nursed his forwards with his usual tantalising collection of tactical kicks and after Kevin Flynn pulled back a try that Tom Kiernan converted for Ireland, another penalty by Price three minutes from time brought Wales her first Triple Crown for 13 years.

First replacement referee in a Welsh Test

The Welsh Triple Crown win over Ireland in March 1965 set up a potential Grand Slam clincher against France in Paris a fortnight later. But in an extraordinary reversal of fortunes, hitherto winless France ran up a remarkable 13–0 lead in the first half-hour against the unbeaten Welsh. Even more surprises were to follow three minutes later when Mr Gilliland, the Irish referee, twisted his ankle and had to leave the field. It meant that for the first time in Five Nations history, a replacement official had to take control of the whistle. The replacement was Bernard Marie of France, a well-respected official in his native France but practically unknown in Britain. Monsieur Marie officiated with absolute rectitude and Wales were thankful to regain some momentum before finally losing 22–13.

58

STAT ZONE

Replacement referees

Wales have been involved in three Test matches when the starting referee has had to be replaced:

Fixture	Year	Ref replaced	Replacement ref
France v Wales	1965	Ron Gilliland (Ire)	Bernard Marie (Fra)
England v Wales	1970	Robert Calmet (Fra)	"Johnny" Johnson (Eng)
Scotland v Wales	2003	Pablo Deluca (Arg)	Tony Spreadbury (Eng)

FAMOUS CHARACTERS

The King

Successful sides need a player who can take a tight game by the scruff of the neck and turn it in their favour. The one and only Barry John filled that role in the team that set out to dominate the Five Nations in 1969. His game was near-perfect. Defenders were left for dead as he jinked, side-stepped or subtly changed pace to drift through heavy traffic. His scores against Scotland and England in the 1969 Triple Crown season were typical John efforts. He was a No 10 who exhibited ghost-like qualities as he spirited his way to the line, quite unlike the quicksilver fly halves Cliff Morgan, David Watkins and Phil Bennett who came before and after him.

John was also an exquisite kicker. He had the ability to land a tactical punt on a sixpence and later in his career became the leading Welsh points scorer of all time (90) thanks to his no-nonsense round-the-corner kicking. An unlikely aspect of his game helped Wales to the 1971 Grand Slam in the showdown match with France in Paris. John would be the first to admit that tackling wasn't his strength, but in taking out big Benoît Dauga as France threatened to score he underlined his bravery. Typically, he complemented his defensive part by making a classic fly half break for the try that paved the way to victory. Like Cliff Morgan before him, in 1972 he left his followers wanting more when, aged only 27, he suddenly retired.

60

The master of the shimmy

There was something of his sharpness off the mark, stabbing side-steps that turned defences inside out or exceptional pace about most of the tries Gerald Davies scored for Wales in a career that spanned Welsh rugby's honour-laden years between 1966 and 1978. When he finally retired, he was the most-capped Welsh threequarter (46) and shared with Gareth Edwards the national record for most Test tries (20).

He began his career in the centre and his initial tries for Wales came in the extraordinary "Keith Jarrett" match against England in 1967. First, he showed his blistering pace to race clear for a try under the posts. Then the famous Davies shimmy created the space for him to sprint diagonally for the corner to finish the scoring. Arguably his most memorable deeds, however, came after transferring to the wing on Wales's tour of New Zealand and Australia in 1969. After taking a sabbatical from representative rugby to concentrate on his studies at Cambridge, he accelerated back into the Welsh side for the 1971 Grand Slam year, opening his new account with two more tries against England at Cardiff and scoring the vital late try at Murrayfield against Scotland. For seven seasons as a wing he helped Wales hold sway over the Five Nations. When Davies was in possession, the threat was obvious; but even without the ball opponents needed to train several pairs of defenders' eyes on his wing. His like was not seen in Wales until Shane Williams entered the national side at the beginning of the millennium.

61

FAMOUS CHARACTERS

The best in the world

The best rugby player in the world was how Gareth Edwards was described in the seventies. He played in 53 successive Tests for Wales between 1967 and 1978, a record that spoke volumes for his strength and resilience, and during his career Wales won three Grand Slams, five Triple Crowns, five outright Five Nations Championships and two shared titles.

The assault on his Test match Everest began in 1967 as a teenager in Paris and he established base camp the next season when he began his fruitful partnership with Barry John. "You throw them and I'll catch them," was John's comment on the slightly wayward Edwards service when they first met. But the scrum half worked to improve his passing and in 1969 was a key member of the Wales team that won the Triple Crown. After a marvellous season in 1971 that included a Welsh Grand Slam and a first Lions Test series victory in New Zealand, Edwards finally scaled the heights when John retired in 1972. He stayed there for six outstandingly successful seasons. Assuming the mantle of tactical controller, Edwards was the linchpin in a side that lost only five times in 24 Five Nations matches and, when he retired in 1978, he held the then record for most career tries for Wales (20), having earlier in his career become Wales's youngest-ever Test captain (aged 20).

Record points on debut

The Wallaby tour to Europe in the winter of 1966/67 was to have far-reaching consequences regarding the future pattern of matches in the Five Nations. The Australians had three of their international matches arranged for January, so to avoid fixture congestion the Home Unions arranged for the Wales–England match to be played in mid-April. It was the first time since 1898 that Wales and England did not meet in January, the weather was absolutely perfect for running rugby and on a firm Cardiff pitch a teenager named Keith Jarrett was catapulted into Welsh folklore, scoring 19 points to set the record points haul for a Welsh player on debut. Wales romped to a sensational 34–21 win that deprived England of the Triple Crown. Jarrett's haul also equalled the Welsh individual match record set by Jack Bancroft against France in 1910 but the killer stat from his drama-filled debut was that his try was the first by a Welsh full back for 33 years. The match was, moreover, only his second first-class appearance in the position. He was a goal-kicking centre for Newport when the selectors turned to him in their hour of need after Terry Price and Grahame Hodgson's form had dipped during a dire run of Welsh defeats against Australia, Scotland, Ireland and France earlier in the season.

63

FAMOUS FIRSTS

Wales's first replacement player

Until the summer of 1968, replacing players in rugby matches was not permitted under International Board regulations. The game's ruling body finally conceded that teams losing a player through injury during the course of an international match should be allowed to replace him, provided medical examination confirmed that the injured player could not continue. The first replacement called on by Wales during a Test was Phil Bennett of Llanelli against France at Stade Colombes in Paris in 1969 when Gerald Davies was injured.

Phil Bennett: It was Norman Gale's alertness that won me my first cap in 1969, against France. We had both been selected as substitutes, and as the game progressed we had resigned ourselves to a quick shower, when suddenly Gerald in the dying minutes of the game was injured and didn't look as if he could carry on. The call to action came so unexpectedly that in my fidgeting excitement the zip on my tracksuit trouser became stuck. There was little I could do about it, but pull at the damn thing and listen to Norman shout at me "Get on quick, there are only a few seconds to go." Suddenly he lunged at my tracksuit bottom, tore it to bits and there I was trotting out on the Stade Colombes pitch for my very first cap and into the history books as Wales's first-ever substitute. I think during those dying minutes Barry had the ball once and he hoofed it, without ceremony, down the field. I hadn't touched the ball in my first international; it had all passed in a flash, and I was still panting after the frustration of battling against my tracksuit zip.

The most-capped subs

International teams were originally allowed to use only two replacements from a bank of four on the bench. The rules concerning bench use, however, have evolved considerably in the past 50 years so that today teams nominate eight players for reserve duties and can use them either as replacements for injured or concussed players, or as straight tactical substitutes. The five most-used Welsh substitute/replacement players are:

Caps off the bench	Player	Career span
36	Gethin Jenkins	2002–2016
32	Ken Owens	2011–2019
32	Dwayne Peel	2001–2011
31	James Hook	2006–2015
31	Paul James	2003–2016

FAMOUS FIRSTS

The super sub

Arguably the most significant replacement appearance made for Wales was that by Ray "Chico" Hopkins against England at Twickenham in 1970. With Wales trailing 6–13 with only 20 minutes of the match remaining, Hopkins came on as replacement scrum half when the Welsh captain, Gareth Edwards, was forced to retire with a leg injury. The bustling Maesteg player revived Welsh hopes, and then crossed for the try which, when converted, gave Wales the lead. Hopkins's try was the first to be scored by a substitute and awarded by a substitute referee. "Johnny" Johnson, the English touch-judge, had been pressed into controlling the match when it was discovered at half-time that the French referee, Robert Calmet, had dislocated his shoulder and broken a bone in his leg in a first-half collision.

66

Home run

Between March 1968 and March 1982 Wales enjoyed a record home run of 27 Five Nations matches without defeat in their capital city. After losing 14–9 to France at the Arms Park in their last match of the 1968 campaign, the dragon roared through 26 wins and a draw in Championship matches at their Cardiff citadel until Scotland silenced Wales with a 34–18 win on the last Five Nations weekend of 1982.

67

FAMOUS CHARACTERS

The indestructible JPR

When the International Board changed the law relating to direct kicking to touch in 1968, JPR Williams was the first of a new breed of full back to effectively exploit the new attacking opportunities offered. He went on to become the master of the full back incursion to the threequarter line, made the position in the Welsh side his own for a decade and eventually retired from the Test side in 1981 as his country's leading cap winner.

Apart from his flair in attack, he was remembered for his rock-solid defence, the first requisite of a full back. He relished making blockbusting tackles and when the ball was in the air, it seemed to be attracted magnet-like to his bread-basket arms. He never flinched and rarely mis-fielded. Indeed, he so relished the contact nature of rugby football that, when injuries dictated, he once started a tour Test for Wales in Australia as a flanker. He left his admirers in Wales a host of happy memories. His indestructible spirit helped his country to three Grand Slams and six Triple Crowns, including four in a row between 1976 and 1979, when he was captain. Playing against the English brought the best out in him. Tries by full backs were uncommon before 1969, yet JPR scored five against them. Moreover, he was on the winning side in all 11 of his appearances against the old enemy.

STAT ZONE

Try-scoring full backs

The first Welsh full back to score a try was Vivian Jenkins, against Ireland at Swansea in 1934 – Wales's 163rd Test match. The full back was regarded first and foremost as the last line of defence. Clean catching, accurate line kicking and rock-solid tackling were the prime requisites for the position. Jenkins had all of these qualities in spades, but his try sparked off a debate: was it the "done" thing for a side's custodian to be up with the threequarters scoring tries? The answer lies in the stats. The next full back to cross for a three-pointer for Wales was Keith Jarrett on his debut against England 33 years later, and it wasn't until JPR's advent and the introduction of the new laws restricting direct kicking to touch, that the duties of the full back changed. Since then, tries by #15s have increased like the proverbial mustard seed. The all-time top five Welsh full back try-scorers:

Full back	Tries for Wales as #15	Career span as #15
Kevin Morgan	12	1997–2007
Lee Byrne	9	2005–2011
JPR Williams	6	1969–1981
Gareth Thomas	6	2003–2007
Leigh Halfpenny	6	2011–2018

Byrne, Thomas and Halfpenny also started – and scored tries for Wales – as wings, but these are excluded from this record.

FAMOUS CHARACTERS

Merve the Swerve

Lean almost to the point of being skinny, Mervyn Davies dominated the back of the Welsh scrum and line-out in 38 successive appearances between 1969 and 1976. An ever present in the winning Lions Test series of 1971 and 1974, he ended his career as the world's most capped international in his position. He was a thoroughbred No 8 at a time when the position was often occupied by converted locks or makeshift flankers, and his relish for weaving runs in open play earned him the nickname "Merve the Swerve".

Davies did not need the trademark jet black hair tightly wrapped in a white sweatband to catch attention on the field. He played by instinct at a time when other No 8s played by numbers. He had an uncanny sense of anticipation in loose play, and with the ball in his hand displayed all the dextrous skills of a top-class basketball player. In defence he cut opponents down with uncompromising tackles. His entry to the Welsh side coincided with the start of a long period of Welsh domination of the Five Nations Championship. Davies featured in two Grand Slam and three Triple Crown sides and captained a new-look Wales side in 1975 (Championship title) and 1976 (Grand Slam).

Captaincy brought to light hitherto unknown aspects of his character. His confidence as a player increased and his tactical reading of games was close to infallible. Wales lost only once in nine outings under his leadership before he suffered a brain haemorrhage while playing for Swansea in a Cup match. Although he made a good recovery, his rugby career was abruptly cut off in its prime.

70

FAMOUS GAMES

The team of all the talents

WALES 22, ENGLAND 6

National Stadium, Cardiff, 16 January 1971

Viv Jenkins, who as teenager, player and long-standing critic had watched international rugby since the twenties, contended that this was the most outstanding side Wales ever fielded. The efficient system of squad training brought together a team of all the talents that was in its prime in 1971. This match in filthy conditions at Cardiff revealed all the components of Wales's dream team working in harmony. The forwards were awesome, overwhelming an inexperienced England pack and giving Barry John and Gareth Edwards the time and space to exercise tight tactical control on the game. The threequarters worked as a unit to transfer the gilt-edged possession quickly to the wings who scored all three of Wales's tries, and behind them JPR Williams was a rock of steadiness organising the defence.

It was a winning formula. Gerald Davies's exquisite finishing brought him two classic wing tries in the south-east corner of the old Arms Park, both converted by the confident left boot of John Taylor, as Wales established an unassailable 16–3 lead at the interval. In the second half JPR Williams, in his role as occasional place-kicker, landed a penalty, and man-of-the-match Barry John nonchalantly dropped his second goal to complete the scoring for Wales and set them on course for their first Grand Slam since 1952.

FAMOUS FIRST AND LAST

The changing value of the try – part I

Growing dissatisfaction with the number of penalty goals that were deciding post-war international matches led the International Board to finally grasp the nettle and tinker with scoring values. In 1971, at the start of the northern hemisphere season, the try was upgraded from three to four points making it, for the first time in the game's history, the most valuable of all rugby's scoring actions.

Barry John scored the last three-pointer for Wales when, in March 1971, he crossed at Stade Colombes in Paris to complete the scoring in a 9–5 Welsh win that sealed the Grand Slam. The scorer of Wales's first four-point try was JPR Williams, against England at Twickenham in January 1972.

FAMOUS CHARACTERS

The Duke

A master-craftsman in the hooking department, Bobby Windsor brought world class talent to 28 successive Welsh international matches through the triumphs of the late seventies. His sound technique, tightness in the scrums, rucks and mauls, and accurate line-out throwing made him the No 1 hooker in the business. Known as "The Duke", he retained his position in an all-Pontypool Wales front-row with colleagues Graham Price and Charlie Faulkner, their combined valley hardness and lion-hearted dedication providing the Welsh backs with the solid platform to practise the match-winning tactics that brought Wales Five Nations domination in the seventies. As a threesome they played together 19 times for Wales, including two Grand Slam seasons. Windsor's strength in the tight and dynamic loose play – the product of a youth spent playing at full back and fly half – won the approval of the Welsh selectors in the autumn of 1973. He scored a try against Australia on debut with an injury time charge that raised the old National Stadium's roof. Although he never again scored in an international, he unfailingly provided the impetus to the forward drive that saw Wales through their most productive spell in Five Nations history.

FAMOUS FIRSTS

The Viet Gwent

When "Charlie" Faulkner and Graham Price joined Bobby Windsor in the Welsh side against France in Paris in 1975, it established a landmark: Pontypool became the first club to provide the Principality with an entire Test front-row. The trio became part of the furniture in the Welsh set-up of the seventies, starting together regularly between 1975 and 1979, and enjoying lasting fame as the fabled "Viet Gwent" in the stories and songs of the Welsh comic legend Max Boyce.

74

FAMOUS CHARACTERS

A fixture at tight-head

One of the most distinguished forwards to emerge from
the Pontypool pack of the seventies was their tight-head,
Graham Price. He played in his first Welsh trial as a callow
youth of 20 in 1972. Three years later he knew all the tricks
of his mysterious front-row trade, winning his first cap
among six newcomers in an experimental side that faced
France in Paris. Wales achieved a convincing 25–10 win and
Price capped a fairy-tale debut by running the length of the
pitch near the end to score the fifth and final try. Thereafter
he was a fixture in the side, playing 39 successive Tests and
finishing in 1983 as their then most-capped forward. During
Wales's crowning years between 1976 and 1979 the tactics
were to subdue and control up front in the early stages of
matches. Then, having established a sound platform, the
forwards would release the ball for the backs to score a flurry
of late tries. The reason the plan worked so effectively was
that Price and his Pontypool front-row colleagues, Charlie
Faulkner and Bobby Windsor, had sucked the energy out of
their opponents. No one gained the upper hand propping
against Price, a player who never had to resort to underhand

115

tactics to establish control. Even when the winning stopped, Price went on plying his trade with customary dignity for Wales. The try-scoring backs had disappeared, but the fit, mobile Pontypool tight-head went on to meet every fresh challenge with relish.

FAMOUS FIRSTS

The crowning years

The expression Triple Crown was an invention of journalists and first appeared in print in the 1890s, though it was England, in 1882/83, who were the first Home Union to carry off the then mythical trophy by beating all of the other Home Unions in the same season. It was not until the late 1970s, however, that any British or Irish team managed to exercise such a firm grasp of the prize as Wales managed. The men in red won four successive Triple Crowns between 1976 and 1979, a feat that has only been matched once since (by England in the 1990s).

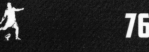

76

The early bath

The Wales–Ireland match at Cardiff on the opening weekend of the 1977 International Championship was the setting for the first dismissals in the history of the ancient tournament. Scotland's Norman Sanson issued marching orders to Geoff Wheel of Wales and Ireland's Willie Duggan for an exchange of punches minutes before half-time. Wheel became the first player to be sent off while playing for Wales in any cap match. To date, eight players have taken the early bath as the result of a sanction while playing for Wales in a Test, three of them at Cardiff and two – Huw Richards and Sam Warburton – in the knockout stages of Rugby World Cup tournaments.

Player sent off	Opponents	Venue	Year
Geoff Wheel	Ireland	Cardiff	1977
Paul Ringer	England	Twickenham	1980
Huw Richards	New Zealand	Brisbane	1987
Kevin Moseley	France	Cardiff	1990
John Davies	England	Cardiff	1995
Garin Jenkins	South Africa	Johannesburg	1995
Sam Warburton	France	Auckland	2011
Ross Moriarty	Argentina	Santa Fe	2018

77

STAT ZONE

The fifty club

Among the many achievements of Gareth Edwards was that of becoming the first Welshman to play 50 Tests for his country. The scrum half who had made his international debut against France in Paris in 1967 reached the milestone in 1978 at Twickenham, piloting Wales to a narrow 9–6 victory with a collection of impressive tactical kicks of all varieties. To put his feat into perspective, it should be remembered that it was not until 1981 that the International Board permitted its member unions to award caps for matches against unions outside the Five Nations, South Africa, New Zealand or Australia. That meant that "internationals" against the likes of Argentina, Italy and Fiji did not qualify for cap status. The first five Welshmen to the fifty cap landmark were:

1st	Gareth Edwards	1978 v England	Twickenham
2nd	JPR Williams	1979 v Ireland	Cardiff
3rd	Robert Jones	1995 v England	Cardiff
4th	Ieuan Evans	1995 v Scotland	Murrayfield
5th	Gareth Llewellyn	1996 v Italy	Rome

FAMOUS GAMES

Winner takes all

WALES 16, FRANCE 7
National Stadium, Cardiff, 18 March 1978

In 1978, for the first time in Five Nations Championship history two sides met at the end of the season with each seeking the Grand Slam, Wales and France playing off in a winner-takes-all finale. In the end, honours went to Wales who achieved their eighth Grand Slam with a performance full of character. As a spectacle, the match had little to commend it. France would have won had they selected an out-and-out place kicker, for between them Jean-Michel Aguirre and Bernard Viviès missed six shots at goal. Even so, Wales deserved their victory. Phil Bennett and Gareth Edwards, each playing his last international for Wales, exercised a tight grip on the game at half-back. Each, moreover, contributed scores at vital times.

France started briskly and were soon ahead through a try by Jean-Claude Skrela from a maul on the Welsh line. Viviès, with a dropped goal, increased the lead to 7–0 before Bennett set the scoreboard moving in Wales's favour in the 30th minute. He attacked on the blind-side before dancing past the French back row for a try which he also converted. In the space of eight minutes, Wales added another seven points. Edwards kicked a towering dropped goal before Bennett was at JJ Williams's left shoulder to take an overhead inside pass and score near the corner to make it 13–7 to Wales at the break. The second half saw Wales defend their lead with grim determination, the only additional score coming in injury time when Steve Fenwick dropped the third goal of the match.

FAMOUS CHARACTERS

A double record-holder

One of the few players who excited spectators during the barren years of Welsh decline in the late 1980s and 1990s was the Llanelli flyer, Ieuan Evans. While Welsh rugby was bereft of leadership off the pitch, Evans was one of the small band of stalwart players whose presence on it meant that there was always hope for a victory. His most priceless asset was his power to beat defenders with unexpected surges of pace. Two particular tries come to mind in this respect. The first, in a breathtaking match against Scotland in the 1988 Triple Crown season, was a spectacular score after he had cut in from the right to beat five defenders on his way to the line. The second was the kick-and-chase that left Rory Underwood standing to bring a rare Welsh victory over England at Cardiff in 1993.

For most of the second half of his international career he was a charismatic national captain whose enthusiasm and engaging personality held together a side of quite limited talent. He became skipper on the eve of the 1991 World Cup campaign and went on to lead Wales a record 28 times that included an unexpected but most welcome Five Nations title in 1994. When he retired from international rugby in 1998 he held the Welsh records for most appearances (72) and most tries (33).

80

STAT ZONE

The dropped goal king

Jonathan Davies was the outstanding Welsh fly half of the late eighties before going north and joining Widnes RL in January 1989. He later, of course, was one of the first players to return to union when it became a professional game after September 1995. He only won 32 union caps for Wales between 1985 and 1997, and it is only a matter of conjecture how Wales might have performed through the lean years of the early nineties if the full panoply of his playing skills and his sharp tactical brain had been available to the national side. Even so, he still managed to set the record for most dropped goals kicked for Wales:

Drops	Player	Caps	Career
13	Jonathan Davies	32	1985–1997
10	Neil Jenkins	87	1991–2002
8	Barry John	25	1966–1972
7	Gareth Davies	21	1978–1985
6	Stephen Jones	104	1998–2011
6	Dan Biggar	70	2008–2019

81

Rugby World Cup record

Wales's best finish in the eight Rugby World Cup tournaments staged to date was third in the inaugural event in 1987. The only other occasion when they finished in the top four was in 2011 when they were beaten by Australia in the bronze play-off. Three times – in 1991, 1995 and 2007 – they have failed to progress beyond the pool stages.

1987 – semi-final exit, losing 49–6 to New Zealand in Brisbane. Won the third/fourth place play-off 22–21 against Australia in Rotorua.

1991 – pool stage exit, losing 16–13 against Western Samoa at Cardiff and 38–3 against Australia, also at Cardiff.

1995 – pool stage exit, losing 34–9 to New Zealand at Ellis Park, Johannesburg, and 24–23 to Ireland at the same venue.

1999 – quarter-final exit, losing 24–9 to Australia at Cardiff (after losing 38–31 in a pool match with Samoa).

2003 – quarter-final exit, losing 28–17 to England in Brisbane.

2007 – pool stage exit, losing 32–20 to Australia at Cardiff and 38–34 to Fiji in Nantes.

2011 – semi-final exit, losing 9–8 to France at Eden Park, Auckland. Lost the third/fourth place play-off 21–18 against Australia, also in Auckland.

2015 – quarter-final exit, losing 23–19 to South Africa at Twickenham.

FAMOUS CHARACTERS

The record points-scorer

An inauspicious start to his Test career in 1991, playing in the first Welsh side for 28 years to lose against England at Cardiff, did nothing to dampen the enthusiasm of Neil Jenkins. The fly half from Pontypridd retained his place and steadily built an excellent reputation as the deadliest place kicker in Welsh rugby history. A chief accountant was needed to keep tabs on his points scoring. The boy from Pontypridd became the most prolific Welsh scorer in Tests, rarely missing with a place-kick offered from anywhere inside his own half. When he addressed the ball for a kick, his body language bore the mark of a confident player expecting to add to his side's total. One lost track of the number of Tests Wales won through his kicking. His contribution of 48 points in the 1994 Five Nations was a significant factor in Wales's only Championship title of the nineties, and he scored all his side's points in the hard-fought victory in Dublin that year. He was also Graham Henry's most important player during the glorious ten-win run that lifted the hopes of the nation in 1999.

Jenkins became indispensable and went on to overtake Ieuan Evans as his country's most-capped player. Even when he was out of favour as a fly half, the selectors could not leave him out of their side. He won a record 70 caps for Wales at No 10, but also played in the centre and at full back, the position he filled in the Test series for the successful 1997 Lions in South Africa.

STAT ZONE

The points machine I

It speaks volumes for Jenkins's prolific scoring that, nearly 20 years since he played his last match for Wales, he remains head and shoulders above the rest of Wales's points-scorers at Test level. The top five Welsh marksmen are:

Career points	Player	Tests	Span
1,049	Neil Jenkins	87	1991–2002
917	Stephen Jones	104	1998–2011
713	Leigh Halfpenny	80	2008–2018
352	James Hook	81	2006–2015
344	Dan Biggar	70	2008–2019

STAT ZONE

The points machine II

Four of the top five positions for most points in a Test match for Wales are also occupied by Jenkins:

Match points	Player	Opponents	Venue	Year
30	Neil Jenkins	Italy	Treviso	1999
29	Neil Jenkins	France	Cardiff	1999
28	Neil Jenkins	Canada	Cardiff	1999
28	Neil Jenkins	France	Paris	2001
28	Gavin Henson	Japan	Cardiff	2004

85

FAMOUS FIRST AND LAST

The changing value of the try – part II

The International Board uplifted the value of a try from four to five points in the middle of 1992. The Swansea flanker Richard Webster scored the last four-pointer for Wales when, in March 1992, he crossed for the only try of the match in Wales's 15–12 win against Scotland. The scorer of Wales's first five-point try was their skipper Ieuan Evans, winning a thrilling chase against England's Rory Underwood after a kick through by Emyr Lewis at Cardiff in 1993. The significance of that score was that it was the first where the five points made a difference to the result: Wales won by the slender margin of one point. The game would have been a draw under the previous season's scoring values.

STAT ZONE

Highest scores and record margins

Wales's failure to progress beyond the pool stages of the 1991 Rugby World Cup condemned them to a grand tour of Europe in 1994, playing against second-tier nations Portugal, Spain and Romania in their bid to qualify for the 1995 event in South Africa. In their first match they registered their highest Test score to date, topping a ton of points for the only time in their playing history to win 102–11 in Lisbon. That, however, is not Wales's biggest margin: the class of 2004 claim that particular distinction with their 98–0 victory against Japan in Cardiff. Wales's biggest winning margins are:

Margin	Score	Opponents	Venue	Year
98	98–0	Japan	Cardiff	2004
91	102–11	Portugal	Lisbon	1994
74	77–3	United States	Hartford	2005
74	81–7	Namibia	New Plymouth	2011
72	81–9	Romania	Cardiff	2001

87

Wales's longest-serving international player

Martyn Williams will be famously remembered as winning exactly 100 caps for Wales. He made his Test debut in a cap match against the Barbarians at Cardiff in August 1996 and quickly became a fixture in the side, packing down on the open-side of the scrum. In August 2011, as captain, he won his 99th Welsh cap, playing against Argentina in a 28–13 win at Cardiff – but was subsequently overlooked for the 2011 Rugby World Cup and was omitted from the squad for the 2012 Six Nations. It looked as if the record books would show him beleaguered on those 99 caps for posterity, but in June 2012 he was named in the squad to play the Barbarians in Cardiff prior to Wales's tour of Australia. He came off the bench to raise his century of appearances 15 years and 283 days since his debut against the same opponents. He was never capped again, but no player has a longer record of international service to the Principality.

STAT ZONE

Record defeats

Welsh rugby reached its nadir in Pretoria in June 1998 when the national side lost 13–96 against the reigning Rugby World Cup champions, South Africa. On the darkest day in the Principality's rugby history, Wales shipped a record number of points (96) and crashed to their biggest losing margin (83 points). The five worst Welsh defeats are:

Margin	Score	Opponents	Venue	Year
83	13–96	South Africa	Pretoria	1998
57	6–63	Australia	Brisbane	1991
57	5–62	England	Twickenham	2007
52	3–55	New Zealand	Hamilton	2003
51	0–51	France	Wembley	1998

FAMOUS GAMES

Opening the new stadium

WALES 29, SOUTH AFRICA 19
Millennium Stadium, 26 June 1999

After two years on the road while the National Ground on the ancient Cardiff Arms Park site was being re-developed, Welsh rugby came home to its new Millennium Stadium in June 1999 and marked the occasion by registering its historic first win against South Africa in 13 attempts. The incomplete stadium had accommodation for 27,000, though in years to come no doubt 270,000 will claim that they were there to witness a famous win.

Wales bristled with purpose from the kick-off. Neil Jenkins kicked them into a 12–6 lead before Mark Taylor broke through to score the first try in the new stadium, just before the interval. Jenkins converted to make it 19–6. A try by Gareth Thomas midway through the second spell effectively wrapped the game up for Wales. Jenkins converted and added a penalty to put Wales 29–14 ahead before Percy Montgomery went over for a late consolation try for the Springboks.

The win consolidated the rising stock of new Wales coach

Graham Henry. Installed at the start of the season, he was to preside over ten successive victories in 1999 – including wins over France (in Paris), England, Argentina (three times) as well as this first one over South Africa.

FAMOUS CHARACTERS

The Great Entertainer

Shane Williams brought a breath of fresh air to international rugby. For the first time since the professional era dawned, here was a player who turned the clock back to the days when a diminutive wing with the wit and innate skills to unlock rigid defences won his right to international honours in a game that had become ruled by giants.

Early in his international career, Welsh coach Graham Henry told the 5ft 7in wing that his physical disadvantages precluded him from substantial Test honours. But the wizard from the Amman Valley whose side-step left opponents grasping thin air focused on building his strength and concentrated on improving his defensive skills. The hours of practice reaped dividends when, in 2003 and after Graham Henry had departed the Welsh scene, Henry's successor, Steve Hansen, showed faith in the wing by recalling him for the Rugby World Cup in Australia. It was Williams's first national call-up for three years, but he never looked back after that. His searing pace injected new life into the Welsh attack in their tournament matches with England and New Zealand, but more significantly cemented his place in the

Welsh starting fifteen for years to come. Ever-present in the 2005 and 2008 Welsh Grand Slams, his try against France in the final match of the 2008 Six Nations was his 41st for the Principality, setting a new career record for a Welsh player. The Welsh nation showed its affection when the great entertainer retired after the match against Australia three years later. Everyone was willing him to score and he duly obliged with virtually the last act of the match at Cardiff, crossing for his 58th try for Wales with a trademark skip and somersault.

91

The top try-scorers

The most likely candidate to wrestle the Welsh try-scoring record from Shane Williams is his former team-mate George North. The leading try-getters for Wales in cap matches are:

Career tries	Player	Tests	Span
58	Shane Williams	87	2000–2011
40	Gareth Thomas	100	1995–2007
36	George North	83	2010–2019
33	Ieuan Evans	72	1987–1998
22	Colin Charvis	94	1996–2007

Charvis is the only forward in the leading five.

92

Wales's oldest try-scorer

When Wales beat Romania 81–9 at Cardiff in September 2001, they notched up 11 tries, including one by Allan Bateman. He had made his Wales debut in 1990 before switching to rugby league, but returned to the union fold when the game went open in 1995. He was 36 years and 197 days old on the day he scored that last of his tries for Wales, making him the oldest Welsh try-scorer to date.

93

The unique dropped goal

All told, Wales have scored more than 120 dropped goals in their
700-plus Tests since 1881. Only one, however, has been kicked
by a forward: that was by flanker Martyn Williams during a
cameo appearance off the bench against Tonga in Canberra
during the pool stages of the 2003 Rugby World Cup.

94

Getting the monkey off their backs

WALES 32, IRELAND 20
Millennium Stadium, 19 March 2005

The 1978 Welsh Grand Slam was the last for a long, long time. The barren years between the decade of the dragon and 2005 had seen Wales challenge just twice for a Grand Slam: in 1988, when Bleddyn Bowen's side was beaten at the final hurdle by France in Cardiff, and in 1994, when Ieuan Evans's team lost 15–8 to England at Twickenham. Moreover, Wales had failed to beat Ireland in Cardiff for 22 years when the men in green stood between Michael Owen's team and its first Six Nations Grand Slam in March 2005. But a side packed with talent and imaginatively coached by Mike Ruddock finally got the monkey off the back of Welsh rugby. In a display that revived the rugby interest of a nation that had wondered if past glories would ever be restored, Wales established an early lead that they never looked like relinquishing.

Gavin Henson and Stephen Jones kicked goals that kept the scoreboard ticking and the flying Welsh backs eclipsed their Irish counterparts. Dwayne Peel revelled in the wealth of possession provided by Robert Sidoli, Ryan Jones and Owen, and the tight forwards were more streetwise in the loose play than their leaden opponents. Indeed, it was the Welsh prop Gethin Jenkins who elicited the loudest cheer of a noisy afternoon when he charged down a poor clearance by Ronan O'Gara and kicked the ball ahead for the opening try of the match after 16 minutes. An earlier dropped goal by Gavin Henson and one of his trademark booming penalties, together with one from shorter range by the reliable Stephen Jones, gave Wales a 16–6 cushion at the interval.

Shane Williams, full of elastic energy, was a constant menace to the Irish and two more Jones penalties put Wales comfortably in the saddle before the *coup de grace* was delivered on the hour. Owen forced a turnover and made ground before teeing up a burst by Tom Shanklin who, finding Kevin Morgan at his shoulder, timed the scoring pass for the full back to perfection. Jones converted and added a late penalty to herald the start of the biggest Welsh celebrations for 27 years.

95

The Hundred Club

Gareth Thomas's last Test for Wales was the infamous 34–38 defeat by Fiji at Nantes which saw the Principality crash out of the 2007 Rugby World Cup at the pool stage. The skipper, however, passed an historic landmark that day, becoming the first player to appear 100 times for Wales. There are only five members of this exclusive club:

Caps	Name	Career span
129	Gethin Jenkins	2002–2016
125	Alun Wyn Jones	2006–2019
104	Stephen Jones	1998–2011
100	Gareth Thomas	1995–2007
100	Martyn Williams	1996–2012

STAT ZONE

Babes in arms

When Tom Prydie lined up on the right wing to face Italy in Wales's final match of the 2010 Six Nations, he broke Norman Biggs's 121-year-old record as the youngest Welsh debutant. The ten youngest players to appear for Wales are:

Player	Age on debut	Opposition	Year
Tom Prydie	18y 25d	Italy	2010
Norman Biggs	18y 49d	NZ Native team	1888
Dafydd Howells	18y 78d	Japan	2013
Evan Williams	18y 213d	England	1925
George North	18y 214d	South Africa	2010
Harry Bowen	18y 226d	England	1882
Tom Pearson	18y 238d	England	1891
Lewis Jones	18y 285d	England	1950
Frank Mills	18y 290d	England	1892
William Thomas	18y 294d	Scotland	1885

The next youngest among the half-a-dozen other Welshmen capped as 18-year-olds is Keith Jarrett, who was a month older than William Thomas when he made his famous debut against England at Cardiff in 1967.

STAT ZONE

Wales's youngest try-scorers

Tom Prydie also holds the record as the youngest to score a try for Wales in a cap match. Only 77 days after his first international appearance, he crossed for a score in Wales's 31–34 defeat by South Africa at Cardiff. He was 18 years and 102 days old then, 136 days younger than the previous Welsh record-holder, Tom Pearson (on his debut against England in 1891). George North is the youngest Welshman to score tries as a new cap. He was 18 years and 214 days old when he crossed twice on his Test debut, against South Africa at Cardiff in 2010.

STAT ZONE

Warren Gatland's Grand Slam record

"If we win our first game, then we can win the Six Nations," Warren Gatland forecast at the start of 2019, his last year in charge of the Welsh national side. It looked a forlorn hope at half-time in the opening game. Wales trailed France 16–0 and seemed destined for oblivion, but a feature of Welsh teams built by Gatland was an ability to stage second-half recoveries. On this occasion they had to thank George North for expertly grasping two opportunities to score tries, while the arrival of Dan Biggar off the bench – so often Wales's go-to man in times of difficulty – was instrumental in saving the day with a kicking master-class that propelled the Principality to an unlikely 24–19 win. What an impact player he proved to be in the big games. Ten changes were made to the starting line-up for the match against Italy seven days later. Keen to replicate the conditions Wales would face at the Rugby World Cup in Japan later in the year, the coach took his squad off to prepare in the south of France before throwing a shadow fifteen into the Roman amphitheatre of the Stadio Olimpico where his rookies delivered a 26–15 victory that was a triumph for the young threequarters Josh Adams and Owen Watkin.

All roads then led to Cardiff and the showdown with Six Nations favourites England. For an hour Wales appeared to struggle against a well-drilled fifteen who led 10–3 at the break and 13–9 after an hour. It was at that stage that Gatland pulled his master-stroke, swapping Gareth Anscombe at #10 for Biggar. The Northampton fly half reprised his Paris role and duly turned the game Wales's way. A patient 30-plus phase passage of play demonstrated the efficiency and accuracy of this Welsh team and culminated in lock Cory Hill ploughing over for a try which put Wales ahead for the first time and raised the roof of the Principality Stadium. Then a late Biggar kick-pass of pinpoint accuracy was juggled by Adams who re-gathered the ball to cross for the try that sealed a 21–13 victory, placing Wales in firm contention for the Grand Slam and taking them to their 12th successive Test win, passing the previous record set between 1907 and 1910.

Scotland next and a visit loomed to Murrayfield, so often the graveyard of Welsh hopes in the past. The Scots have become a difficult side to beat at home in recent years but Adams, Jonathan Davies and Hadleigh Parkes were prominent in the first half and helped Wales establish a 15–6 half-time lead. For most of the second stanza, however, Scotland dominated possession and territory, and only the patient defensive shield devised by Gatland's wily assistant coach Shaun Edwards prevented the hosts scoring more than one try. At the death Anscombe landed a penalty, but the 18–11 winning margin rather flattered Wales, though arguably the chaos precipitated by Project Reset which in the week before the match proposed a regional merger of the Ospreys and Scarlets did little to settle the mindsets of many of the players.

And so back to Cardiff for a tilt at the Grand Slam against Ireland, the reigning Grand Slam holders. There were mind games aplenty beforehand. Awful weather was forecast for the match yet Ireland's coach Joe Schmidt asked for the roof to be retracted. The game was played at the mercy of the elements but Wales showed that they were masters of the wind and rain unleashed on their hallowed Cardiff acre. The side that traditionally started slowly went ahead after a minute when Anscombe delicately chipped over the advancing Irish threequarters for Parkes to gather and score. Anscombe converted before having to switch to full back early on when super-sub Biggar replaced the injured George North in a re-arranged Welsh back division. But the steady New Zealander kept his calm, polished his sights and carried on kicking goals in the teeming rain to put Wales 16–0 ahead at the interval. He successfully landed three more penalties in the second half and by the 70-minute mark Wales were sitting on an unassailable 25–0 lead. Only a late converted consolation try blotted the copybook, but 25–7 was a win beyond the dreams of most of Wales's delirious followers, and it brought Warren Gatland his record third Grand Slam: no other head coach had achieved more than two in Five or Six Nations history.

The Welsh pack were immense throughout the season. Among Warren Gatland's many skills were shrewd man-management and an ability to pick team leaders who were outstanding motivators, the very heartbeat of their teams. His earlier Grand Slam forward generals, Ryan Jones (in 2008) and Sam Warburton (in 2012 – as well as at the 2011 Rugby World Cup and in wins against England to take the 2013 Six Nations and 2015 Rugby World Cup match at Twickenham), had played highly significant roles in Gatland's previous successes. The champion of Wales in 2019 was another born pack leader in Alun Wyn Jones, fittingly the only survivor of Gatland's first Welsh fifteen against England back in 2008. Time and again in 2019 Alun Wyn provided the inspiration, technical expertise and emotional intelligence to extract the best from his troops. He was the obvious choice as the Guinness Man of the Tournament at the end of the campaign, and one former player and leading critic spoke for the entire Welsh nation when he described Jones as "the greatest Welsh rugby player of the professional era".

99

THE HEAD HONCHO

Welsh head coaches

The silver lining to the cloud over Wales's humiliating 3–24 defeat by South Africa in 1964 – their worst Test reverse for 40 years – was the resulting root and branch change to how the game was run in the Principality. The upshot was that Wales led a coaching revolution in the Home Unions. Ray Williams was installed as overall coaching organiser in 1967 and later the same year the Welsh Rugby Union appointed the former No 8 David Nash as their national team coach. His first assignment was the international against the All Blacks in Cardiff in November 1967, and since then every Welsh side to take the field for a Test has enjoyed the benefit of coaching input.

Head coach	Tenure	P	W	D	L	Win rate
David Nash	1967–68	5	1	1	3	20%
Clive Rowlands	1968–74	29	18	4	7	62
John Dawes	1974–79	24	18	0	6	75
John Lloyd	1980–82	14	6	0	8	43
John Bevan	1982–85	15	7	1	7	47
Tony Gray	1985–88	18	9	0	9	50
John Ryan	1988–90	9	2	0	7	22
Ron Waldron	1990–91	10	2	1	7	20

Head coach	Tenure	P	W	D	L	Win rate
Alan Davies	1991–95	35	18	0	17	51
Alex Evans	1995	4	1	0	3	25
Kevin Bowring	1995–98	29	15	0	14	52
Dennis John	1998	2	1	0	1	50
Graham Henry	1998–2002	34	20	1	13	59
Lynn Howells	2001	2	2	0	0	100
Steve Hansen	2002–04	29	10	0	19	34
Mike Ruddock	2004–06	20	13	0	7	65
Scott Johnson	2006	3	0	1	2	0
Gareth Jenkins	2006–07	20	6	1	13	30
Nigel Davies	2007	1	0	0	1	0
Warren Gatland	2008–19	114	64	2	48	56
Robin McBryde	2009–17	6	5	0	1	83
Rob Howley	2012–17	20	10	0	10	50

RUGBY IN THE BLOOD

Down the years Welsh rugby families have provided the Principality with an impressive set of examples of fathers, sons and brothers who have turned out in international matches.

Grandfather and grandson

Dai Hiddlestone (first capped in 1922) and Terry Price (first capped in 1965)

Fathers and sons

Tom (first capped in 1882) and Paul (first capped in 1921) **Baker-Jones**
Len (1951) and Roger (1974) **Blyth**
Idwal (1939) and Brian (1962) **Davies**
Nigel (1988) and Sam (2016) **Davies**
Jack (1924) and Billy (1947) **Gore**
Howell (1904) and Howie (1930) **Jones**
Windsor (1926) and Geoff Windsor (1960) **Lewis**

Paul (1986) and Ross (2015) **Moriarty**
Derek (1972) and Scott (1993) and Craig (1995) **Quinnell**
Jim (1970) and Tom (2001) **Shanklin**
Glyn (1912) and Rees (1947) **Stephens**
George (1903) and Bill (1937) **Travers**
Brynmor (1978) and Lloyd (2011) **Williams**
Dai (1987) and Thomas (2017) **Young**

Kuli (first capped by Tonga in 1988) was the father of Taulupe
(first capped by Wales in 2011) **Faletau**
George (first capped by England in 1905) was the father of
Walter (first capped by Wales in 1938) **Vickery**

Stepfather and stepson

Paul (1980) and Jamie (2001) **Ringer**

Brothers

Billy (first capped in 1890) and Jack (first capped in 1909)
Bancroft
Jack (1929) and Arthur (1934) **Bassett**
Norman (1888) and Selwyn (1895) **Biggs**
Nathan (2003) and Aled (2007) **Brew**
Terry (1953) and Len (1954) **Davies**
Jonathan (2009) and James (2018) **Davies**
Tom (1898) and George (1900) **Dobson**
Charlie (1906) and Louis (1910) **Dyke**
Bob (1882), Arthur (1885) and Bert (1892) **Gould**

Dai (1882) and Bill (1884) **Gwynn**

George (1881) and Theo (1888) **Harding**

Bertie (1912) and Tom (1927) **Hollingdale**

Evan (1890) and Dai (1891) **James**

Willie (1925) and Tommy (1935) **James**

Bobby (1926) and Dick (1929) **Jones**

David (1907), Jack (1908) and Tuan (1913) **Jones**

Gareth (1989) and Glyn (1990) **Llewellyn**

Teddy (1902) and Willie (1910) **Morgan**

Charles (1891) and David (1894) **Nicholl**

Syd (1888) and Gwyn (1896) **Nicholls**

Andy (1995) and Steve (1997) **Moore**

Richard (1981) and Paul (1986) **Moriarty**

Tom (1919) and Dai (1924) **Parker**

Wick (1920) and Jack (1923) **Powell**

Glyn (1934) and Dai (1934) **Prosser**

Scott (1993) and Craig (1995) **Quinnell**

John (1927) and Bill (1929) **Roberts**

Jamie (2001) and Nicky (2003) **Robinson**

Dai (1891) and John (1891) **Samuel**

Harold (1936) and David (1937) **Thomas**

Bernard (1925) and Maurice (1933) **Turnbull**

Harry (1912) and Jack (1914) **Wetter**

Bleddyn (1947) and Lloyd (1957) **Williams**

Gareth (1980) and Owain (1990) **Williams**

Richard (1988) and Matthew (1996) **Wintle**

Matt (first capped for Australia in 1997) was the brother of
Brent (first capped for Wales in 2003) **Cockbain**

Frank (first capped for Wales in 1885) was the brother of
Froude (first capped for England in 1886) **Hancock**

STAT ZONE

The international record

Up to 31st March 2019, Wales had played 722 international matches, won 379, drawn 29 and lost 314.

Opponents	First cap match	P	W	D	L
Argentina	1991	18	13	0	5
Australia	1908	42	11	1	30
Barbarians	1990	4	2	0	2
Canada	1987	12	11	0	1
England	1881	132	58	12	62
Fiji	1985	11	9	1	1
France	1908	97	50	3	44
Georgia	2017	1	1	0	0
Ireland	1882	127	69	7	51
Italy	1994	27	24	1	2
Japan	1993	10	9	0	1
Namibia	1990	4	4	0	0
New Zealand	1905	34	3	0	31
New Zealand Natives	1888	1	1	0	0
New Zealand Army	1919	1	0	0	1
Pacific Islands	2006	1	1	0	0
Portugal	1994	1	1	0	0
Romania	1983	8	6	0	2

Samoa	1986	10	6	0	4
Scotland	1883	125	73	3	49
South Africa	1906	35	6	1	28
Spain	1994	1	1	0	0
Tonga	1986	9	9	0	0
United States	1987	7	7	0	0
Uruguay	2015	1	1	0	0
Zimbabwe	1993	3	3	0	0
TOTAL	**1881**	**722**	**379**	**29**	**314**

The record against Australia includes the 1927 match against New South Wales subsequently given full Test status by the Australian Rugby Union

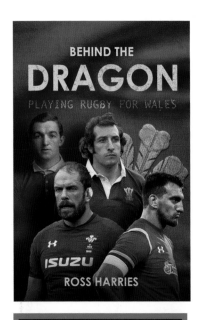
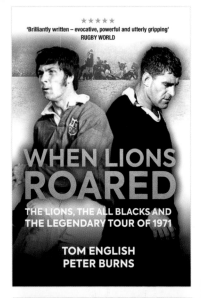
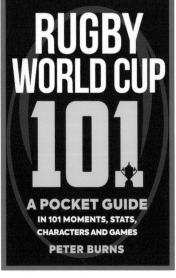
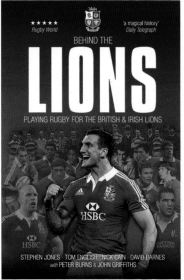